TRADITION AND INDIVIDUALITY

SYNTHESE LIBRARY

STUDIES IN EPISTEMOLOGY,

LOGIC, METHODOLOGY, AND PHILOSOPHY OF SCIENCE

Managing Editor:

JAAKKO HINTIKKA, *Boston University*

Editors:

DONALD DAVIDSON, *University of California, Berkeley*
GABRIËL NUCHELMANS, *University of Leyden*
WESLEY C. SALMON, *University of Pittsburgh*

VOLUME 221

J. C. NYÍRI

Institute of Philosophy, Hungarian Academy of Sciences

TRADITION

AND

INDIVIDUALITY

Essays

KLUWER ACADEMIC PUBLISHERS
DORDRECHT / BOSTON / LONDON

Library of Congress Cataloging-in-Publication Data

Nyíri, János Kristóf.
    Tradition and individuality : essays / J.C. Nyíri.
        p.   cm. -- (Synthese library ; v. 221)
    Includes bibliographical references and index.
    ISBN 0-7923-1566-9 (alk. paper)
    1. Tradition (Philosophy) 2. Wittgenstein, Ludwig, 1889-1951.
3. Knowledge, Theory of. 4. Language and languages--Philosophy.
I. Title. II. Series.
B105.T7N95 1992
148--dc20   91-39683

ISBN 0-7923-1566-9

Published by Kluwer Academic Publishers,
P.O. Box 17, 3300 AA Dordrecht, The Netherlands.

Kluwer Academic Publishers incorporates
the publishing programmes of
D. Reidel, Martinus Nijhoff, Dr W. Junk and MTP Press.

Sold and distributed in the U.S.A. and Canada
by Kluwer Academic Publishers,
101 Philip Drive, Norwell, MA 02061, U.S.A.

In all other countries, sold and distributed
by Kluwer Academic Publishers,
P.O. Box 322, 3300 AH Dordrecht, The Netherlands.

printed on acid-free paper

For Ilonka

TABLE OF CONTENTS

PREFACE   ix

CHAPTER 1: Wittgenstein's New Traditionalism   1

CHAPTER 2: Wittgenstein 1929–31: Conservatism and Jewishness   9

CHAPTER 3: Collective Reason: Roots of a Sociological Theory
of Knowledge   25

CHAPTER 4: Some Marxian Themes in the Age of Information   39

CHAPTER 5: Tradition and Practical Knowledge   47

CHAPTER 6: "Tradition" and Related Terms: A Semantic Survey   61

CHAPTER 7: Historical Consciousness in the Computer Age   75

CHAPTER 8: On Esperanto: Usage and Contrivance in Language   85

CHAPTER 9: Heidegger and Wittgenstein   93

CHAPTER 10: Writing and the Private Language Argument   105

NOTES   115

BIBLIOGRAPHY   159

INDEX   171

PREFACE

During the last fifteen years I have experimented with two successive approaches to the problem of the social embeddedness of individual behaviour. The first approach was community-centred, traditionalist; the second more individualistic. The present volume contains essays from both phases: chapters one to six representing the earlier, chapters seven to ten the later approach.

I have chosen to have those earlier essays published along with the later ones for two reasons. First, because my current arguments (for what they are worth) clearly presuppose the background of my former endeavours. Secondly, I still do not think that my earlier attempts were entirely misdirected. In fact the choice in favour of the one or the other position depends on the evaluation of a single cluster of arguments: those pertaining to the epistemological significance of *writing*. I have come to believe that it is the medium of writing which creates, historically and psychologically, a space for individual, critical thinking. But one can have doubts about how far this space extends; how deeply it permeates social interactions; whether the written word will retain, in the future, the role it has now. And to the extent that communication remains pre-literal, or becomes post-literal, the issue of traditionalism still appears to be a living one.

I am fully aware that the transformations in my views are not independent of recent political changes in Eastern Europe; that both my earlier and later theoretical attitudes are in some measure reflections of a certain political bias. By presenting them side by side, I hope to avoid the danger of apparent extremism which either alone might represent. The source of my traditionalist leanings was the Hungarian experience of social decay, from the 1960s to the 1980s: the recognition that the Hungarians have become a people without traditions, that their traditions were undermined by a totalitarian bureaucracy, and the impression that they were yearning not just for freedom, but also for rootedness, stable institutions, social cohesion. Today the politically dominant ideology in Hungary is a traditionalist one, and the fruits it bears seem to be hypocrisy, intolerance, social and economic bankruptcy. It seems that stable institutions in a modern society do, after all, rely on a liberal culture; on learning, education, and critical thinking.

I wish to acknowledge the profound debt I owe to Professor Georg Henrik von Wright for essential inspiration and assistance. His influence upon my work has been decisive. Research fellowships granted by the Alexander von Humboldt Stiftung in 1986–87 and 1990–91 for work at the University of Bochum gave me the leisure and opportunity to develop my ideas in detail, and to write most of these essays. I am greatly indebted to my colleagues in Bochum. Special thanks are due to my host Prof. Gert König, who not only provided generous scientific guidance, but also convinced me to put together this volume and was instrumental in organizing its publication. To my friend Barry Smith I am indebted for the invaluable and unfailing support he gave me during all these years. Without his stimulus and constant help the volume could not have been completed.

The essays have been to some extent edited in order to avoid overlaps. "Wittgenstein's New Traditionalism" appeared originally in *Essays on Wittgenstein in Honour of G. H. von Wright* (*Acta Philosophica Fennica* 28/1–3, 1976), pp. 503–512. In this essay the reference to Wittgenstein's "General Remarks", a manuscript compilation since published as *Vermischte Bemerkungen*, could of course have been easily updated. I left it as it is deliberately, in order to preserve something of the flavour of those early days of scholarly work on Wittgenstein's manuscripts. "Wittgenstein 1929–31: Conservatism and Jewishness" is an abridged version (with note 25 newly added) of the study "Wittgenstein 1929–1931: Die Rückkehr" (KODIKAS/CODE – *Ars Semeiotica* 4–5/2, 1982, pp. 115–136, English translation in Stuart Shanker, ed., *Ludwig Wittgenstein: Critical Assessments*, vol.4, London: Croom Helm, 1986, pp. 29–59). I wish to express my thanks to the members of the now sadly defunct Wittgenstein Archive in the University of Tübingen, without whose technical support this study, prepared in October 1979, could not have been written. "Collective Reason: Roots of a Sociological Theory of Knowledge" appeared in W. Gombocz *et al.*, eds., *Traditionen und Perspektiven der analytischen Philosophie: Festschrift für Rudolf Haller*, Wien: Hölder-Pichler-Tempsky, 1989, pp. 600–618, note 48 newly added. "Some Marxian Themes in the Age of Information" appeared in *Doxa* 15 (Budapest: 1989), pp. 169–182. "Tradition and Practical Knowledge" appeared in J.C. Nyíri and B. Smith, eds., *Practical Knowledge: Outlines of a Theory of Traditions and Skills*, Beckenham: Croom Helm, 1988, pp.

17–52; the permission of Routledge (formerly Croom Helm) to reprint this piece is gratefully acknowledged. "'Tradition' and Related Terms: A Semantic Survey" appeared in *Doxa 14 – Semiotische Berichte* 1–2/1988 (Budapest – Vienna), pp. 113–134. "On Esperanto: Usage and Contrivance in Language" appeared in Rudolf Haller, ed., *Wittgenstein – Towards a Re-Evaluation*, Wien: Hölder-Pichler-Tempsky, 1990, vol. II, pp. 303–310. The permission of the publisher to reprint this essay, and the essay "Collective Reason", is gratefully acknowledged. Chapters 7, 9 and 10 have not been hitherto published.

The volume is dedicated to my wife.

Dunabogdány, June 1991
J. C. Nyíri

CHAPTER 1

WITTGENSTEIN'S NEW TRADITIONALISM*

Towards the end of his life Wittgenstein wrote: "Men have judged that a king can make rain; *we* say this contradicts all experience. Today they judge that aeroplanes and the radio etc. are means for the closer contact of peoples and the spread of culture".[1] This remark is a rather clear allusion to what Wittgenstein, in my opinion, always believed: that man's so-called historical progress, and especially the positive role reason is supposed to play in it, is an illusion. The same conviction is reflected in Wittgenstein's choice of the motto for his *Philosophical Investigations*, a quotation from Nestroy: "Überhaupt hat der Fortschritt das an sich, daß er viel größer ausschaut, als er wirklich ist". That this motto refers to the social-historical progress of mankind and not, say, to Wittgenstein's own progress in philosophy, becomes obvious when viewed together with the Foreword written to the *Philosophical Remarks*, dated 6.11.1930, where Wittgenstein states that the spirit of his work is different from that of the mainstream of European and American civilization, since the latter is characterized, as the former is not, by the idea of a constant progress. Wittgenstein's attitude towards the liberal idea of progress is that of a conservative. This attitude is, actually, not conspicuous in the *Philosophical Remarks*, to which the Foreword was written subsequently; but it was already there, I believe, at the time of the *Tractatus* and it becomes quite manifest in Wittgenstein's later writings. My purpose in the present paper is to suggest that Wittgenstein's so-called later philosophy is the embodiment of a conservative-traditionalist view of history, and, in particular, to show that this philosophy in fact provides a logical foundation for such a view.

There can be no doubt at all that the conservative-traditionalist sentiment really was a basic element in Wittgenstein's personality. True, many of his later friends in England were left-wing; and his plans, in the thirties, to settle down in the Soviet Union, might suggest communist sympathies. But Wittgenstein could have been attracted by the personal integrity of some Marxists without being in agreement with their political views; and even if he shared, as I believe he did, their contempt towards bourgeois society, the background of that contempt was, in Wittgenstein's

1

case, not at all leftist. Indeed, when Wittgenstein said that our age was a dark one,[2] the standard by which he thus judged was not the vision of some future, ideal community; his sympathies were with past, patriarchal, authoritarian societies; and if it was not just the Russia of Tolstoy and Dostoevsky that he wished to rediscover in the USSR, then what appealed to him must simply have been the puritanical and authoritarian character of the Soviet way of life.[3] His friend during and after the first World War, Paul Engelmann speaks of his loyalty towards all legitimate, *genuine* authority, whether religious or social; an attitude so much his second nature that revolutionary convictions of whatever kind appeared to him throughout his life as simply *immoral*. Wittgenstein, who, as Engelmann writes, saw through and abhorred all *unjustified* conventions, became through this attitude "a figure utterly beyond the comprehension of the 'educated' of our day".[4] And Fania Pascal, Wittgenstein's Russian teacher in Cambridge in the mid-thirties, writes that "at a time when intellectual Cambridge was turning Left he was still an old-time conservative of the late Austro-Hungarian Empire".[5] As it faded away, pre-war Austria became for Wittgenstein a standard, in comparison with which the decades following the war represented a sunken, miserable age.[6] One can, in Wittgenstein's case, very clearly observe the process, described by the sociologist Karl Mannheim, in the course of which conservatism becomes a theory – the process, in which "an die Stelle eines schlichten Lebens aus einem alten Lebenskeime ein Haben der alten Lebensform auf der Ebene der Reflexion, auf der 'Ebene der Erinnerung' tritt".[7]

The theoretical crystallization of Wittgenstein's intuitive and frustrated traditionalism seems definitely to have been furthered by the impact he received from Spengler's work *The Decline of the West*. In the years 1931–33 Spengler's name is mentioned more than once in Wittgenstein's writings,[8] and this seems to be the very period during which the essential elements of Wittgenstein's later philosophy first emerge. In the *Philosophical Grammar* many important ideas of the *Philosophical Investigations* are anticipated; indeed, many of the remarks contained in the earlier work appear practically unaltered in the later one. On the other hand, the *Philosophical Grammar* is quite unlike the *Philosophical Remarks*. The stylistic difference between the two works[9] is really striking – and this difference reflects a profound philosophical re-orientation. It has been suggested that the cause of this re-orientation was that famous paper by Gödel, on formally undecidable propositions of *Principia Mathematica*, published in 1931.[10] But it seems that Wittgenstein was not acquainted with that paper before 1935.[11] Spengler's influence, on the other hand, is quite obvious. What was significant for Wittgenstein in his philosophical re-orientation was not only Spengler's general pessimism as regards the fate of Western culture and his description of the type of man characteristic of the present age,[12] but also the idea that Western culture and its specific mode of thinking is just one among many, that the Western spirit is, since the beginning of "modern times", in a process of decay, and that it is *Russianism* that today represents "Spring" in contrast to the "Winter" of the "Faustian" (Western) nations and their culture which by now has degenerated into a "civilization". Wittgenstein, too, speaks early in 1931 of our "half-rotten culture", and of Russia, where "passion" still promises something, against which our "idle talk" is powerless. These expressions occur in a conversation with Schlick.[13] In a previous discussion Wittgenstein comes to discuss Schlick's views on ethics. Schlick wrote that there were, in theological ethics, two views as to the essence of the Good: according to the shallower interpretation the Good is good because God so wills it; according to the profounder interpratation God wills the Good because it is good. Wittgenstein disagrees with Schlick. He believes that the first interpretation is the more profound one: good is, what God commands.[14] Values, customs, traditions cannot and should not be *explained*.[15] Every explanation is, as it were, a judgement of reason – but reason itself, as Wittgenstein in his later philosophy sets out to prove, is, in the last analysis, grounded in traditions. It is in this sense that Wittgenstein will constantly emphasize that all explanation must be done away with and description alone must take its place; that what has to be accepted, the given – is not explanations, but *forms of life.*

The results of Wittgenstein's later philosophy can be summed up by saying that "freedom", if this should mean something different from being bound by genuine traditions, is simply incompatible with reason; and that "nonconformism" is an anthropological absurdity. Wittgenstein nowhere explicitly formulates these results. The principle he stated in his early work, according to which "whereof one cannot speak, thereof one must be silent", is the one he now *practices*. His analyses deal with concrete, almost banal, phenomena of human behaviour. The key concepts of modern European philosophy – concepts like freedom, individuality, autonomy – Wittgenstein does not use. Those concepts belong to a certain *picture* of man, they are bound up with certain *presuppositions* – and Wittgenstein wants to liberate both his readers and himself from that picture and from those presuppositions entirely. Alice Ambrose describes the lectures Wittgenstein held in 1932–33. "He used the language of everyday speech", she reports. "And there was no hint of mysticism, no reference to the unsayable. What was puzzling was his use of picturesque examples, which in themselves were easily comprehensible, but of which the *point* they were intended to make escaped one. It was like hearing a parable without being able to draw the moral".[16]

Now one such parable, often told by Wittgenstein, runs as follows. I give someone the order to fetch me a red flower from the nearby meadow. How is he to know what sort of flower to bring, when I have only given him a *word*? How will he get from the colourless word – to a red flower? Well, one is inclined to say that he carries a red image in his mind as he sets out to look for a red flower, and he compares it with the flowers to see which of them has the colour of the image. Wittgenstein finds many faults with this answer. He suggests, first of all, that one should try to describe the procedure without recurring to the vague concept of a red mental image – since what a mental image achieves can obviously be accomplished just as well with the help of, say, a red bit of paper. The following will then take place: our man carries a chart with him on which names of colours are co-ordinated with coloured squares. When he hears the order "fetch me a red flower", he draws his finger across the chart from the word "red" to a certain square, and goes and looks for a flower which has the same colour as the square. Now there certainly is such a way of searching, says Wittgenstein, if for instance one looks for an unusual shade of some colour – but it isn't the only or the usual way. Usually we go, look about us, walk up to a flower and pick it, without comparing it to anything. And the actual problem – how does one get from the word to a flower? – is not, in any case, solved by the above-offered explanation. For how does one get from the *spoken* word "red" to the *written* name of that colour in the chart? We could again, of course, imagine some sort of chart that should now guide us. "But in fact", writes Wittgenstein, "there isn't one, there's no act of memory, or anything else, which acts as an intermediary between the written sign and the sound".[17] In the last analysis, a sign is followed by the action without there being any kind of mediation between them. "Consider the order", suggests Wittgenstein in one of his most striking examples, "'*imagine* a red patch'. You are not tempted in this case to think that *before* obeying you must have imagined a red patch to serve you as a pattern for the red patch which you were ordered to imagine".[18] One is, as a child, says Wittgenstein, *trained* to *react* in a certain way to colour words and to words generally; and the success of the training, the fact that every normal person reacts to colour words in the *same* way, makes in this respect understanding and communication possible at all.

The moral of the above examples (and of the innumerable similar ones Wittgenstein constructed between 1931 and 1936) is a negative one. Our actions, in the last analysis, are not guided by reflection; and successful communication must be regarded, so to speak, as a miracle. Liberal anthropology, based on the enlightened conception of man as guided by reason, thus turns out to be quite untenable. Now the negative observations Wittgenstein makes are not, at first, paralleled, in his writings, by positive suggestions as to a traditionalist anthropology. From 1937 on, however, the outlines of such an anthropology gradually emerge. Some main steps in this process of theoretical development are represented by Wittgenstein's use of the term *Unerbittlichkeit* (inexorability) from 1937 on, of the term *Gepflogenheit* (custom) from 1943 on, and of the expressions *konform urteilen* (judge in conformity) and *Autoritäten anerkennen* (recognize authorities) in 1950 and 1951. As to the external circumstances in which the transitions here indicated occurred, it may be noted that the years 1937–38 and 1943 represent important stages in Wittgenstein's life. The confession Fania Pascal speaks about[19] reflects, in 1937, a psychological crisis, and 1938 is the year of the *Anschluß*, of Wittgenstein's application for British citizenship and of his plan to publish the existing version of the *Philosophical Investigations*; 1943 being the year in which that plan was considered for the second time.

The application of a word – e.g. the word "red" – consists in *following* a certain *rule*. Indeed the phenomenon of following a rule plays an altogether central part in our thinking and acting. One follows grammatical rules when speaking; logical ones when reasoning; mathematical ones when calculating. What does it actually mean, however: to follow a rule? Well, one is *guided* by the rule; as by an order. There seems to be a gulf, however, between any order and its performance. "When we give an order", writes Wittgenstein in 1933,

it can look as if the ultimate thing sought by the order had to remain unexpressed, as there is always a gulf between an order and its execution. Say I want someone to make a particular movement, say to raise his arm. To make it quite clear, I do the movement. This picture seems unambiguous till we ask: how does he know that *he is to make that movement?* – How does he know at all what use he is to make of the signs I give him, whatever they are? – Perhaps I shall now try to supplement the order by means of further signs, by pointing from myself to him, making encouraging gestures, etc. Here it looks as if the order were beginning to stammer.[20]

A sign-post, an arrow for instance, can be regarded as a simple rule[21] – but even the sign-post begins to stammer, if one keeps asking oneself: in *which* direction should I then go – how should I *interpret* that arrow? At every step, suggests Wittgenstein in a remark written in 1936, different *decisions* seem to be possible,[22] since a rule does not determine its own application. "Then can whatever I do be brought into accord with the rule?" By 1944, Wittgenstein is ready to reject this question.

Let me ask this: what has the expression of a rule – say a sign-post – got to do with my actions? What sort of connexion is there here? – Well, perhaps this one: I have been trained to react to this sign in a particular way, and now I do so react to it. – But that is only to give a causal connexion; to tell how it has come about that we now go by the sign-post; not what this going-by-the-sign really consists in. On the contrary; I have further indicated that a person goes by a sign-post only in so far as there exists a regular use of sign-posts, a custom.[23]

The correct application of a sign is the application upon which people agree; this very fact of agreement has to be recorded, indeed demanded,[24] if we want to grasp the *possibility* of linguistic actions or, for that matter, thinking, at all. The phenomenon of language rests on the regularity of, on the agreement in, behaviour. This agreement must be something quite fundamental, certainly not of the kind one can argue about – it is, rather, the basis of any argument, any discussion. "If language is to be a means of communication there must be agreement not only in definitions but also (queer as this may sound) in judgements".[25] Even in order to err, writes Wittgenstein in one of his last aphorisms, a man must already *judge in conformity* with mankind.[26]

It is here, in the remarks written during the last two years of his life, that Wittgenstein's traditionalism, and, in particular, the close connection of this traditionalism to his later theories, becomes most apparent. One must, writes Wittgenstein, "recognize certain authorities to make judgments at all";[27] authorities, for instance, like our *school*,[28] or our inherited world-picture;[29] *foundations*,[30] against which any doubt is hollow.[31] "My *life*", writes Wittgenstein, "consists in my being content to accept many things".[32] One must, in particular, be content to accept inherited *language-games*. The language-game is there – like our life.[33] The thesis that language-games, i.e. forms of life, have to be accepted, that they are what is *given*, is of course there in the *Investigations* too, and is indeed, from 1931 on, implicit throughout Wittgenstein's writings. In any endeavour to criticize a given linguistic tradition, only another linguistic tradition can serve as a standard. Language, then, cannot be subjected to criticism from the standpoint of "pure" thinking. Languages come into being and become obsolete, and different linguistic traditions become interconnected, exerting a soft pressure on each other. Important thereby is that a language must always be something that has *grown organically*[34]; formalized languages are merely adjuncts, extensions of everyday language. "Our language", writes Wittgenstein in that well-known passage, "can be seen as an ancient city; a maze of little streets and squares, of old and new houses, and of houses with additions of various periods; and this surrounded by a multitude of new boroughs with straight regular streets and uniform houses".[35]

Lovingly rests Wittgenstein's gaze upon this natural manifoldness of language. He does not search for something "common" behind the diversity of linguistic phenomena, he does not search for the "essence" of language, with reference to which one could introduce, as it were, a new linguistic order. Our language – and the time it reflects – suffers, it is true, from many illnesses, and Wittgenstein certainly does not want to *conserve* "the darkness of this time".[36] But the sickness of a time, says Wittgenstein in a famous dictum, is cured by an alteration in the mode of human life – and that alteration is brought about by some cause or other, as the result of some development or other; not, however, through a medicine that an individual could purposely invent.[37]

CHAPTER 2

WITTGENSTEIN 1929–31:
CONSERVATISM AND JEWISHNESS

When, at the beginning of 1929, Wittgenstein resolved to live for a time once more in Cambridge and to concern himself again with philosophical problems, the outlines of the thoughts which he was to develop gradually over the next twenty years were by no means clear to him. Although the *Weltanschauung*, the general attitude which pervades his later reflections is clearly present in the manuscript notes that he made at this time, there is here almost no connection between the elements of this general disposition and theoretical argument, no interplay between attitudes and concept-formation. Wittgenstein grappled with problems in the first few months of 1929 in a manner which appears – in the light of what he was later able to achieve – as a directionless wandering about. He was indeed often conscious of this, and it filled him with despair. "Once more in Cambridge. How strange. It sometimes appears to me", he wrote on the 2nd of February in his notebook, "as if time had been turned back. ... I don't know what is awaiting me. But something will turn up! If the spirit does not leave me. ... The time here should have been – or should be – in fact a preparation for something. I have to become clear about something" (MS 105, p. 2). And a few days later: "Everything that I am now writing in philosophy is more or less insipid stuff. But I still believe it possible that it should get better" (*ibid.*). "I should like to know", he wrote in another remark dating from the Spring of 1929, "whether this work is right for me. I'm interested in it, but not inspired. ... Somehow I see my present work as provisional. As a means to an end" (MS 106, p. 4). Or again: "I am continually moving in circles around the problem. Apparently without ever coming any nearer to it" (MS 106, p. 30). Several months pass; Wittgenstein feels himself still to be pursued by doubts, by feelings of directionlessness, uncertainty, which come to expression in a typical manner for example in the dream which he describes in a passage of the 6th of October, immediately followed by the remark: "I'm disgruntled because my work gets me no further. Emptyheadedness" (MS 107, p. 154).

Many of the questions with which Wittgenstein concerned himself at the beginning of 1929 reveal, of course, an obvious continuity both with

9

problems of the *Tractatus* as also with the fundamental themes of his later writings. These are, above all, certain questions in the foundations of mathematics, Wittgenstein's interest in which – as is well known – had been re-awakened by Brouwer's lecture in Vienna in March 1928. And it is to considerations in the foundations of mathematics which, for example, the following important remark is directed:

I believe that the mathematics of the last century had experienced a period quite peculiarly lacking in instinct, from which it will suffer for a good deal longer. This instinctlessness is connected, I believe, with the decline of the arts, it flows from the same cause (MS 106, p. 253).

It would be wrong, however, to conceive considerations in the philosophy of mathematics as the essential driving force of Wittgenstein's thinking in this period. On the contrary: he finds himself "thrown back on problems of arithmetic" as it were "against his will" (MS 105, p. 19). He sees in arithmetic an "unconquered fortress of the enemy": with this enemy in one's rear one cannot "march into the territory of psychology" (MS 107, p. 39) – and it is precisely this territory that seems to have held Wittgenstein's interest. Thus on the 9th of October he writes in his notebook: "I am conscious that the most magnificent problems are lying in my closest vicinity. But I can't see them, or I can't grasp them" (MS 107, p. 156). And a similar entry from the day following: "I feel today an unusual poverty of problems around me; a sure sign that there lie *before me* the most important and the hardest of problems" (MS 107, pp. 158 f.). His "Freudian resistance to finding the truth" (MS 107, p. 100) seems to have loosened itself only gradually.

The theoretical path followed by Wittgenstein in 1929 was a slow and uncertain one. The next two years however brought decisive insights. Already on the 3th of January 1930 he is confronting the "naive conception of the meaning of a word", which would have it that "in hearing or reading a word one 'presents', 'imagines' [*vorstellt*] its meaning to oneself" (MS 108, p. 61, cf. *Philosophical Remarks* [*PR*], § 12). His first comparison of the question "what is a word" with the question "what is a chess-piece" appears, it seems, on the 15th of January (MS 107, p. 240, cf. *PR* § 18), and already on the 19th of May he is talking of "*grammatische Spielregeln*". "Different kinds of chess-pieces: bishops, knights, etc., correspond", he argues, "to different kinds of words" (MS 108, p. 169). And on the next day:

I have hit here upon that method of explaining signs which Frege derided so much.

One could, that is to say, explain words like "knight", "bishop", and so on, by giving the rules which relate to these pieces (MS 108, p. 170).

This new conception of the meaning of a word must, of course, bring with it a new conception of believing, thinking, and so on. *Thinking* is, according to a passage from the 29th of June, "the use of symbols" (MS 108, p. 201), and "the thought – if one can talk of such a thing at all – must be something totally familiar" (MS 108, p. 216, entry of July 19). The thought is nothing "ethereal" (*ibid.*), nothing "amorphous". And it is fundamentally something that can be "observed by everybody". "One could express this as follows". Wittgenstein writes, "that in the *thought* there is nothing private" (MS 108, p. 279, entry of July 31). The sense in which the thought must be conceived as something non-private is outlined particularly vividly in a passage from the 25th of August:

If I were to resolve (in my thoughts) to say "abracadabra" instead of "red", how would it show itself that "abracadabra" stood in place of "red"? How is the position of a word determined? Supposing that I were to replace all the words of my language simultaneously by others, how could *I* know which word stood in place of which other word? Is it here the ideas [*Vorstellungen*] that remain and hold fixed the positions of the words? As if there were a sort of hook attached to each idea, upon which I hang a word, which would indicate the position? This I can't believe. I cannot make myself think that ideas have a place in understanding different from that of words (MS 109, pp. 45 f.).

In a strikingly short time, practically in the last days of July 1930, there take shape also those stylistic peculiarities which are so characteristic of Wittgenstein's later writings: the dialogue and unanswered question, the familiar "*Du*" as a form of address. Each of these had occasionally appeared already in his earlier notes (for example in a passage of January 3, MS 108, p. 56), but it is nevertheless not until the end of July that they become a regular stylistic device. Thus on the 29th of July:

"Yes, *that* is what I expected". How could you have expected it, when it wasn't yet there at all? (This misunderstanding contains the entire problem of our reflections and also its solution) (MS 108, pp. 265 f.).

The following passage derives from the 31st of July:

"I thought to myself, he will now come". – "Yes, you said 'he is about to come', but how do I know that you meant that by what you said?" (MS 108, p. 274)

Now one can, after all, ask: "How, then, does it show itself, that he means the picture as a portrait of N.?" – "Well, in that he says that it is". – "But then how does it show itself that he means that by what he says?" – "In no way at all" (MS 108, p. 275).

This style reflects completely Wittgenstein's theoretical intentions. It must be in the everyday circumstances in which language is used, for example in conversations, that it becomes manifest whether particular philosophical questions or concepts have sense at all. In such situations it will be revealed in a convincing manner that "everything is ... after all simple and familiar to us all" (MS 109, p. 15, entry of August 16), that particular words – for example the word "to mean" – will most readily and naturally allow themselves to be driven back "from their metaphysical to their correct application in the language" (MS 110, p. 34).

If philosophers use a word and inquire into its meaning one has always to ask oneself: is, then, this word as a matter of fact used in this way in the language which has created it // for which it has been created //?
One will then usually discover that it is not so and that the word is used against // in a manner contrary to // its normal grammar. ("Knowledge", "being", "thing") (MS 109, p. 246).

It is not as if the normal – inherited – grammar were somehow capable of being given foundation of its own through special insights; rather it is the foundation of every insight and of every judgement. " 'To understand calculation in the primary school, a child would have to be a great philosopher; failing that, they have to have exercise, training'" (MS 109, p. 138, entry of September 13; cf. *Zettel*, § 703). To say, "this is just how we use language" (MS 109, p. 224) or "this is how I have learned language" (MS 109, p. 286) is to present *ultimate justifications*; the need for further justification has to be conceived as springing from a "misunderstanding of the logic of our language" (MS 109, p. 225). "Teach them to us" – arithmetic or language – "and you have provided them with a foundation" (MS 111, p. 63).

In a similar way the question whether someone "really sees the same colour ... as I" when he sees, for example, blue, also reveals itself as a misunderstanding of the logic of language (MS 109, pp. 298 f.). "Does he really see the same as I when he looks at the sample?" To raise doubts here, Wittgenstein argues, is as nonsensical as the assumption that thoughts are a "secret and blurred process, ... of which we see only indications in language" (MS 109, p. 99). It is only apparently the case

that we cannot know whether two human beings see the same colour when they look at an object. This is nonsense, for by seeing two different colours we mean // understand // something quite different, and in this sense there exist criteria as to whether the two see the same or different colours (MS 109, p. 171).

One is, Wittgenstein stresses, "led into error by a false analogy", "when one says that ideas or images are private" (MS 153, p. 59, transferred into notebook 110 on 6 July 1931). The question: "how do you know that that which you call red is really the same as that which the other calls red?" is "just as nonsensical" as the question "how do you know that that *is* a red spot?" (MS 109, pp. 196 f.). In order to set forth clearly the senselessness of these and similar questions Wittgenstein, already in 1930, employs the method which later, for example in the *Blue Book*, was to play such a fundamental role: he shows that the actual or assumed function of mental pictures can in every case be fulfilled also by physical pictures. One can

substitute for the process of calling to mind images in thought, another process, say the writing down of signs (or some other process), which performs the same task (MS 109, p. 89).

Someone receives the instruction to look for, say, a yellow flower. One might here wish to suppose (and this is *the* inherited philosophical assumption) that in looking for the flower he carries around with him in his memory an image of the colour yellow, comparing it with each successive flower that he sees. This memory image can of course in principle be replaced by a yellow colour-sample, and the question raised by Wittgenstein is: how will this person know which flower is of the *same* colour as the sample?

It is perhaps most instructive to think that, when we look for the flower with a yellow sample in the hand, then at least the relation of colour-similarity is not present to us in a further image. But rather that with this we are quite contented (MS 110, pp. 277 f.).

But if one can do without the image of colour-similarity, then – leaving aside special cases – the sample of colour, and thus also the mental image, are not required either.

I go looking for the yellow flower. And even if whilst I am walking along an image should appear to me, do I really need it, when I see the yellow flower – or any other? (MS 110, p. 276)

And an example which shows clearly that the mental image can in some cases play *no* role as model:

The command is given: "Imagine to yourself a red circle". And I do this. How would it be possible that I follow the words in that way? (MS 110, p. 173, written in March or April 1931)

There must come a point where ideas, models, signs or images, no longer serve as supports for action, where the action is its own support.

Already in 1931 Wittgenstein seems completely to be in possession of this decisive insight. And indeed it is clear that the elements that predominate in Wittgenstein's remarks of 1931 are exactly those which were to constitute the groundwork of his later synthesis. In 1931 there were composed the remarks on Frazer, in which Wittgenstein lays so great an emphasis on the programme of *mere description* (MS 110, p. 180), insisting thereby that philosophy

is not allowed to disturb in any way the actual // factual // use of language // what is actually said //, it can in the end only describe it.

For it cannot justify it either. It leaves everything just as it is (MS 110, pp. 188 f.)

– a conception which would not, of course, be capable of being adopted if word-meanings were *independent* of the use of words, if the latter were determined by the former. But such an independence does not exist: "To understand the meaning of a word means to know or understand how it is used" (MS 111, p. 12). It is in 1931 also that Wittgenstein outlines those examples and arguments which are familiar from the opening sections of the *Philosophische Untersuchungen*: the criticism of Augustine's conception of language (MS 111, pp. 15 ff.), the game with building blocks (MS 111, pp. 16 f.), the rejection of the idea that there is something common to all games (MS 111, p. 17, cf. also especially MS 111, pp. 79 ff., where the concept of *family resemblances*, and shortly thereafter the expression itself, are anticipated). The interconnections between these elements do not, of course, immediately meet the eye. Wittgenstein himself indeed notes on the 14th of October 1931 that what he is saying seems on the one hand to become "ever easier to understand, its significance, on the other hand, is ever more difficult to grasp". And the interpretation of these remarks is impeded still more by the fact that, in the time-span here considered, there is no fixed correlation of problems and concepts – that Wittgenstein is continually changing his terminology. (It is, he wrote on the 29th of August 1930, "as though the problem were moving house", MS 109, p. 67.) Thus for example the series of concepts depiction–verification–application–use retains, between 1929 and 1931, its connection with one and the same problem,[1] but this problem is further developed and modified. And thus also the role of *application* and *action* in relation to *understanding* becomes visibly clarified through the concept "plan" (see e.g. MS 109, pp. 81 ff.) – this concept itself however is soon abandoned. Yet these elements do combine together into a unified whole – if not conceptually, then certainly from the point of view of the underlying general attitude. "Whatever I write", Wittgenstein remarks in 1930, "is fragments, but he who understands will be able to extract from them a self-contained world-view" (MS 108, p. 152). A self-contained world-view indeed: the world-view of conservatism.

WITTGENSTEIN'S CONSERVATISM

A characterization of Wittgenstein's general attitude as "conservative" makes sense only if, and to the extent that, it points to well-defined theoretical and historical parallels or influences. Conservative thinking is, historically, an extremely heterogeneous formation, and in particular the German so-called neo-conservatism of the 1920s and 1930s with which Wittgenstein's later thought can be most readily compared, differs essentially from, say, the first significant wave of German conservatism which occurred at the end of the eighteenth and the beginning of the nineteenth centuries. There are, nevertheless, certain fundamental ideas which are common to both currents, and these ideas are indeed characteristic for most of those theoretical and political movements that have associated themselves, or have been associated, with the term "conservative".[2] That these fundamental traits are all of them present in Wittgenstein's later writings – including the manuscripts of 1929–31 – is unmistakable. The *rejection of the rationalistic scheme of explanation* is a guiding idea not only of the later *On Certainty*,[3] but also of the commentary to Frazer; the *respect for what exists*, for the *historically given*, is expressed not merely in those programmatic remarks which draw attention to the purely descriptive task of philosophy, but in Wittgenstein's analyses in general, which rest, as a matter of principle, upon the acceptance of the authority of everyday language.

Wittgenstein's later writings, beginning with the manuscripts of 1929–31, imply an image of man which stands in glaring contradiction to the enlightened, liberal view. The concept of the internally or mentally autonomous, rational individual, of the human subject acting in accordance with the light of his reason, sovereign within his own mental world, reveals itself as absurd in the face of the realization that the meaning of a word is not a mental image, but the use to which the word is put; thinking, believing, expecting, hoping, and so on, are not private mental processes; mathematical insight is grounded in exercise, in drill;[4] every action is executed, ultimately, without any kind of interpretation of models.

Wittgenstein's conservative anthropology employs predominantly negative formulations: it must move, after all, like all conservative theories, within a system of concepts that is in fact alien, that has been borrowed of necessity from the false world-picture to which it is opposed. Thus it is not for any inexplicable, mystical reason that Wittgenstein stands "in struggle with language" (MS 110, p. 273, cf. *Culture and Value* [*CV*], p. 11), that he must set his hopes on the "inexpressible" (MS 153a, p. 130, cf. *CV*, p. 16). By 1930, however, that which is inexpressible seems to lie more deeply hidden, to be set further back, than was the case in the *Tractatus*. The historical surroundings of the young Wittgenstein had to some extent preserved elements – for example the still living idea of an established order – which could, as it were, simply be depicted, be pointed out, within a conservative theory. The world in which Wittgenstein lived after the War was altogether different: to someone with the dispositions of an Austrian conservative it could not but have appeared entirely alien.

That Wittgenstein was under the immediate influence of some leading neo-conservative figures – Spengler, Dostoevsky, and almost certainly also Moeller van den Bruck – can be easily shown. The ideas by which he was affected were of course put forward already before 1930, even though, for economic and political reasons, it was at just this time that they became most widely disseminated.[5] The expression "conservative revolution" occurs already in 1921, in application to Nietzsche and Russian literature, in a work of Thomas Mann.[6] Dostoevsky's pronouncement – "we are revolutionaries from out of conservatism" – was cited already by Moeller van den Bruck in his introduction to *The Devils* in his German collected edition of Dostoevsky's works.[7] And it seems to be precisely Dostoevskian ideas, as these were given coinage by Moeller, which served as Wittgenstein's most basic introduction to the intellectual world of neo-conservatism.[8] That Wittgenstein's later philosophy shows an utter lack of understanding for the conservative values inherent in Western, middle-class forms of life, that his epistemological traditionalism became entirely prejudiced against European ideals, is, then, hardly independent from the detrimental effect Dostoevsky had upon him. Dostoevsky's counterposition of Russia and the degenerate *civilization of the West* is, of course, a recurrent theme also in the work of Spengler, the most influential of the neo-conservative thinkers of the post-War years.

That Spengler had exerted at just this time (i.e. 1930/31) a quite particular influence upon Wittgenstein can be seen clearly from the passages printed in the *Vermischte Bemerkungen*. In the drafts which he made in 1930 of possible forewords to the text now known as *Philosophische Bemerkungen*, Wittgenstein affirms that where the course of "European and American civilization" tears everything along with it, the "value of the individual" is no longer capable of expressing itself in social institutions and in social actions "as it is in the age of a great culture". Culture, Wittgenstein writes,

is like a great organization, assigning to everyone who belongs to it a place where he can work in the spirit of the whole, and his strength can with much justification be measured in terms of this whole. In times of non-culture energies fritter away and the strength of the individual is worn down by opposing forces and frictional resistances... (MS 109, p. 205, cf. *CV*, p. 6).

There are a number of manifest parallels between certain neo-conservative tendencies of the 20s and 30s and many of Wittgenstein's thoughts in the same period. And it is indeed possible to point to passages where one can speak not merely of parallels but of actual influences upon Wittgenstein's thinking. But the question can be raised as to the extent to which Wittgenstein – as he gradually began to develop the themes and some of the central theses of his later philosophy, and as he found his own characteristic mode of expression – was conscious that, in his theoretical endeavours, he was taking part in a burning contemporary discussion. To what extent was the history of German neo-conservatism in the 1920s and 30s a part of Wittgenstein's own personal fate? The answer to this question can be anticipated in one sentence: Wittgenstein must have been intensely interested in the outcome of at least certain discussions within neo-conservatism – those relating to *the German-Jewish problem*, which at this time both deeply affected Wittgenstein himself, and had a powerful influence upon neo-conservative thinking; those relating to the problems of Jewish character, of Jewish society, and of the relation between Jew and Christian.

CONSERVATISM AND JEWISHNESS

S. M. Bolkosky, in his book *The Distorted Image*, estimates the number of anti-semitic books published in Germany between 1929 and 1932 at over seven hundred, and puts the number of German-Jewish counter-

publications at double this number.[9] Certain publications within this flood
of writings have, of course, an especial significance. One such was the
special issue on "The Jewish Question" of the periodical *Süddeutsche
Monatshefte* which appeared in September 1930 and which included
contributions by both Jewish and anti-semitic authors. One contribution,
by the "conservative revolutionary" Ernst Jünger, bearing the title "On
Nationalism and the Jewish Question", is particularly suited as a summary
introduction to the themes that here concern us. Jünger pokes fun at "the
strange blossomings of well-bred conservative prose which are these days
ever more frequently flowing out from Jewish pens. Bitter declamation in
defence of culture, witty and ironical attacks upon the bustle of civilization, an aristocratic snobbism, the Catholic farce..." The Jew, Jünger
writes, "certainly cannot complain about the attention given to him by
those powers who believe themselves to be the representatives of our
present-day thinking". But this attention is, Jünger believes, misplaced.
The Jew is after all "not the father, but the son of liberalism – just as, in
absolutely everything else having to do with German life, both the good
and the bad, he can play no creative role".[10] These, then, were the issues
of theoretical controversy:
– what role was played by the Jew in the victory of "liberalism", of the
  bourgeois-capitalist social order?
– are Jews able to participate in a true "culture", or rather only in a
  "civilization"?
– does there exist an unbridgeable chasm between Jewish and Christian
  – and especially Catholic – religion and world-view?
– is the Jew able to be truly creative, or is he always merely imitative?
  Richard Wagner, already in the nineteenth century, was able to call
Jewishness [Judentum] the "bad conscience of our modern civilization"[11]
and to affirm:

The Jew in general speaks the modern European languages only as if acquired and not
as if he were a native. This rules out for him any capacity to express himself properly
and independently within them in accordance with his essence. A language, its
expression and its development, is not the work of individuals but of a historical
community: only he who has grown up unconsciously within this community can take
part also in its creations. ... In [our] language, [our] art the Jew can only repeat what
others say, affect the art of others, he cannot compose or create works of art in a
manner that would speak authentically.[12]

Thus Otto Weininger, when he spoke of the "necessary lack of genius
in the Jew", of his "lack of any truly rooted and original conviction",[13]

was only taking up again what had already been often repeated.

Another question which arose repeatedly in the discussions has been mentioned already above. It was the question of the essential, or merely accidental, connection between bourgeois-liberal progress – "civilization" – and Jewishness. This question was of course touched upon also by Spengler. "In the moment when the civilized methods of the European-American world-cities shall have arrived at their full maturity, the destiny of Jewry – at least of the Jewry in our midst (that of Russia is another problem) – will be accomplished".[14] Spengler characterizes the city-dweller as traditionless; it was, however, equally repeatedly affirmed that this lack of reverence for the traditional does not belong to the *essence* of Jewishness. Thus Rudolf Kaulla for example, in his *Der Liberalismus und die deutschen Juden: Das Judentum als konservatives Element* of 1928, wrote that

Form signifies tradition, the preservation of that which obtains. Form belongs to what one calls the "culture" of a people, formlessness something that does not take this culture seriously. Form has an integrating effect, formlessness dissolves. Formlessness nourishes a falling apart. – And it would almost certainly be impossible to find a more vivid illustration of the truth of these propositions, and thereby also – to put things most simply – of the dangers of "modernism" than the Jewish religion and its fate ... having been caught up by the enlightenment, which has in part mitigated and modernized its old forms, in part set them aside.[15]

That Wittgenstein deals strikingly often with the problem of the Jewish mind in the remarks published as the *Vermischte Bemerkungen*, is stressed by G. H. von Wright in his lecture "Wittgenstein in Relation to his Times" which was presented simultaneously with the publication of these *Bemerkungen*, and can indeed be interpreted as an introduction to the remarks which the latter contains. I wish here to enlarge upon von Wright's discussion of this problem by means of an analysis of the material contained in the *Vermischte Bemerkungen* from the period 1929–31, in the light of its wider context in Wittgenstein's manuscript.

The first such passage, which appears on p. 72 of notebook 107, reads:

The tragedy consists in this, that the tree does not bend, but breaks. The tragedy is something non-Jewish. Mendelssohn is probably the most untragic of composers (cf. *CV*, p. 1).

That Wittgenstein is here ascribing to himself the traits which he sees in Mendelssohn is clear, since he adds, immediately after the sentence concerning Mendelssohn, a further sentence in which he talks of his own untragic nature, of his untragic "ideal" (MS 107, p. 72). And indeed only a few manuscript-pages later he writes:

Mendelssohn is like a man who is only jolly when the people he is with are all jolly anyway, or good, when all around him are good, and not really like a tree which stands fast, as it stands, whatever may take place around it. I too am like that and am inclined to be so (MS 107, p. 120, cf. *CV*, p. 2).

Mendelssohn is mentioned by Wittgenstein also in several other places: for example on p. 98 of notebook 107, where he speaks of an "Englishness about him" (cf. *CV*, p. 2), and two years later, in September 1931, on p. 195 of notebook 111, where he writes: "Mendelssohn's music, where it is pefect, is musical arabesque. This is why we have a sense of embarrassment at his every lack of rigour" (cf. *CV*, p. 16). And even though this is not perhaps immediately clear from the quoted lines themselves, both of these remarks refer to the Jewishness in Mendelssohn. Does not Weininger, after all, who was so especially highly regarded by Wittgenstein, speak of the "similarity, to which attention has been drawn since Wagner, between the Englishman and the Jew"?[16] And did not Wagner himself, in his essay "On Jewishness in Music", write that he could feel himself caught up by Mendelssohn only

when there is offered to our phantasy, which seeks only to be more or less entertained, nothing other than the displaying, laying out, and interlacing, of the smoothest and most refined and artistically polished figures, as in the ever-changing stimuli of colour and shape of the kaleidoscope – never, however, where these figures are intended to take the form of deeper and more rigorous sensations of the human heart[17]

– an intention which leads, in Mendelssohn, merely to "dissolute and phantastical shadow-images".[18]

Wittgenstein's last-mentioned remark on Mendelssohn is, at it happens, followed immediately in the manuscript by the passage which follows it in the *Vermischte Bemerkungen*:

In western civilization the Jew is always measured on scales which do not fit him. Many people can see clearly enough that the Greek thinkers were neither philosophers in the western sense nor scientists in the western sense, that the participants in the Olympian Games were not sportsmen and do not fit in to any western occupation. But it is the same with the Jews. And by taking the words of our language as the only possible standards we constantly fail to do them justice. So at one time they are overestimated, at another underestimated. Spengler is right in this connection not to classify Weininger with the philosophers // thinkers // of the West (MS 111, pp. 195 f., cf. MS 153a, p. 122; *CV*, p. 16).

The idea that the Jew is to be measured not by Western but rather precisely by oriental standards had in fact become established already in German intellectual history as a characteristic correction or supplementation to the demand for total emancipation and assimilation (as put forward for example by Lessing). It was defended by the dialect poet and folk-author J. P. Hebel, who was also one amongst Wittgenstein's most favoured writers. Thus in his study "Die Juden" Hebel wrote of the "characterizing mark" "which the climate of the land where the Bible was written has impressed upon its children"[19] and which has, through the centuries, by no means disappeared. The Jews have remained entirely true to this "consecration of their homeland", and thus they have "more character and strength", Hebel believes, that the people of the West.[20] It seems therefore clear that it was not only as an exponent of the idea that "a great part of our lives ... is a – pleasant or unpleasant – stumbling about through words" and that "most of our wars are ... wars of words",[21] but also on the strength of his views on Jewishness, that Hebel may have captured Wittgenstein's interest.

Now the reference to Spengler in the remark analyzed above clearly relates to that passage in the *Untergang des Abendlandes* where Spengler speaks of three Jewish saints of the last centuries – "which can be recognized as such only through the colour-wash of Western thought-forms".[22] He refers, in particular, to Otto Weininger

whose moral dualism is a purely Magian conception and whose death in a spiritual struggle of essential Magian experience is one of the noblest spectacles ever presented by a Late religiosity. Something of the sort Russians may be able to experience, but neither the Classical nor the Faustian soul is capable of it.[23]

The concept of a "Jewish saint" occurs in fact also in Weininger's own work (albeit in a negative sense: "In the Jew, almost as much as in the Woman, good and evil are not differentiated from each other; there is certainly no Jewish murderer, but not either is there such a thing as a Jewish saint",[24] and indeed again in Wittgenstein: "Amongst Jews 'genius' is found only in the holy man. Even the greatest of Jewish thinkers is no more than talented. (Myself for instance)" (MS 154, p. 16; CV, p. 18). These sentences occur at the beginning of that remarkably instructive paragraph in which Wittgenstein speaks of his "merely reproductive" thinking and of "Jewish reproductivity" in general, providing a list of thinkers who had exerted an influence upon him. This paragraph, as it appears in the notebook 154, is connected directly to those remarks by which it is followed also in the *Vermischte Bemerkun-*

gen (*CV*, pp. 18–20). Almost immediately after this series of entries there follows the remark concerning the Jewishness in Rousseau (MS 154, p. 21; *CV*, p. 20), with a further chain of reflections, relating to the history of the Jews in Europe – and including Wittgenstein's reference to the inadequate "rigour" of Mendelssohn.

The cause of this interest on Wittgenstein's part in the Jewish mind and in the peculiarities of the Jewish character seems to have been a deeply personal one. By this is meant not so much the fact of his own partially Jewish extraction, but much rather the circumstance that – to his distress – he believed himself to have detected in his own personality, as it seemed, those traits which had been held in the literature – for example in Weininger's *Sex and Character*, a work whose subject matter was of the deepest personal significance to Wittgenstein – to be precisely characteristic of the Jews. The significance of this problem of his own Jewishness can be gauged, for example, by the dream which he describes in an entry in his notebook of the 1st of December 1929. The central character in this dream is an evil man, who had disowned his Jewish descent. His name is given by Wittgenstein alternatively as "*Vertsagt*" and "*Vertsag*", but is written also as "*Verzagt*" and interpreted by Wittgenstein as "*verzagt*" (disheartened). There is, however, a more obvious interpretation, which is nevertheless avoided by Wittgenstein: that he (who is not of course *versagt*: betrothed) is worried by the fact that as a human being and as a philosopher he has *versagt* (failed), and that it is *versagt* (denied) to him, as a Jew, that he should create a decent and profound work. – After this dream, almost a year goes by before the theme of Jewishness is once again mentioned by Wittgenstein in his writings. In the meantime he has made decisive theoretical advances, and completed the manuscript of a book which, although (or indeed *because*) it has been written by a Jew, is not adapted to "the current of European civilization" (MS 109, p. 206; *CV*, p. 6).

Wittgenstein's draft foreword to this book, from which these words have been extracted, derives from the 6th of November 1930. One day earlier Wittgenstein had entered into his notebooks the passages concerning Renan's *Histoire du Peuple d'Israël* (MS 109, pp. 200–202) which have been published in the *Vermischte Bemerkungen* (*CV*, pp. 5 f.). In the first of these passages there is so much that recalls the commentary to Frazer, that it does not come immediately to light: although Wittgenstein speaks here of primitive man and primitive peoples he is in fact referring to the ancient *Jewish* people. Had he wanted to concern himself simply with primitive peoples and customs he then would certainly not have chosen Renan as his scientific reading matter. His driving motive seems much rather to have been a personal and subjective interest in that which was Jewish – as does indeed become clear from the second passage, which follows immediately after the first in the manuscript also: "What Renan calls the 'bon sens précoce' of the semitic races (an idea which had occurred to me too a long time ago) is their *unpoetic* mentality, which heads straight for what is concrete. This is characteristic of my philosophy".[25] Now, however much Wittgenstein may have found to disagree with in Renan's elucidations, he must nevertheless have found the perspective in which Renan set the Jewish problem to be of profound interest. In the foreword to his book Renan had characterized "the founders of Christianity" as "direct descendants of the prophets",[26] and had acknowledged the opposition between Christianity and the "liberal rationalism of the Greeks": "Christianity will leave behind ineradicably a trace, and liberalism will no longer rule the world alone".[27] "The history of the Jews and of Christianity", he goes on,

have been the joy of a full eighteen centuries, and even though half conquered by Greek rationalism they still possess an astonishing power for ethical betterment. The Bible in its different forms remains, in spite of everything, the great book, the comforter of mankind. It is not impossible that the world, in becoming exhausted by the repeated declarations of the bankruptcy of liberalism will become once more Judeo-Christian...[28]

As had been mentioned already, Wittgenstein's remarks on Renan are then followed by sketches for a foreword – entered into his notebooks on the 6th and 7th of November (*CV*, pp. 6 f.) – in which Wittgenstein distances himself from Western "civilization", and on the 8th of November he writes that most familiar version of the foreword which was published in the *Philosophische Bemerkungen*. Thus his thoughts on the Jewish spirit on the one hand, and his remarks on culture and civilization (i.e. his most directly conservative remarks) on the other hand, are connected inseparably together. And in his manuscripts the topic of Jewishness, around 1930, is bound up in turn with themes – for example with the ideas of the fundamental role of common sense and of concreteness – which will permeate his later writings.

Wittgenstein did not, however, have a comprehensive or coherent conception of Jewish history, or of the Jewish inheritance, the Jewish character and intellect. His references are, in general, impressionistic in form, having no special claims to validity (as when he says, for example, that "the Jew is a wasteland, beneath the thin layer of rock there lie however the fiery-flowing masses of the spirit", MS 153a, p. 161, cf. *CV*, p. 13). It would be mistaken to see in Wittgenstein's later work the incorporation of any definite current of thought that would normally be conceived as traditionally Jewish. Nevertheless Wittgenstein's interest in Jewishness is not merely a psychological or biographical fact. It can, first of all, help to explain the interconnections which obtain between his later work and German neo-conservatism. But more profoundly, considered from the perspective of religious typology, one recognizes that his thought does not simply exhibit strongly *Catholic* traits, as was the case with neo-conservatism in general, but rather precisely those traits which are *common* to Catholic and Jewish thought, but alien to Protestantism, in particular to Lutheranism. It would perhaps be not incorrect to turn back, at this point, to the already mentioned special issue of the *Süddeutsche Monatshefte*. Here the fundamental characteristics of the Jewish religion were presented by Leo Baeck. This religion is, Baeck wrote,

a religion of commandment and of the deed. ... The word, even the word of confession, and the expression of faith in general, has less weight within it than does action.

To speak of God

is only to attempt to make the inexpressible capable of expression. This ultimate futility is sensed with such an intensity that one covers over with silence the ancient word for the eternal God. For him who seeks to find his way on this earth, it is only the deed that fulfils God's command, that becomes a manifestation of Him.

Jewish religiousness is a "religiousness of the deed", and "wherever a Jewish community has preserved the old forms of life", there exist

manifold customs and practices, extending into the most minute, of which he who perceives them from the outside must suppose that they conceal and strangle religion, and of which he who possesses and practises them can learn that they consecrate everyday life.[29]

It is known, however, that the Catholic Church too declares faith to be insufficient, and accordingly makes that which is good dependent not upon faith alone, but upon its becoming proved through deeds – where the Protestant conception recognizes as good works only those deeds which, as is said, flow from out of the living faith. Wittgenstein thus shows, in fact, that the Protestant, in particular Lutheran, conception *must* be false. It is impossible, he shows, to speak meaningfully of intentions, of purposes, of willing and believing – outside, that is, a context of deeds, customs and practices.

CHAPTER 3

COLLECTIVE REASON:
ROOTS OF A SOCIOLOGICAL THEORY OF KNOWLEDGE

1. In a number of influential papers on the history of the Vienna Circle, Rudolf Haller has repeatedly drawn attention to the fact that the received view of this history is distorted. In particular he has argued that the notions entertained within the Circle encompassed much of what is usually attributed to today's *critics* of the logical positivists, and that Neurath and Carnap developed ideas which anticipated the theory of science of T. S. Kuhn. In line with Hempel's 1935 interpretation,[1] Haller points to the holistic character of Carnap's and Neurath's epistemology and to the historical and sociological orientation displayed especially by the latter. Neurath the sociologist, writes Haller,

fully realized that scientific research is the process of production of a collective that is determined by its social-economic situation and by history, especially its own history. ... Why this collective decides in favour of any given conception amongst the alternatives, why it chooses this rather than some other version of the correspondence of sentence and system, this, according to Neurath, cannot be given further justification. And thereby we stand at just that point at which Kuhn's theory appears as an immediate continuation of that of Neurath.[2]

Now holism and sociologism are of course different perspectives: Neurath's precursors Mach, Poincaré and Duhem were holists, but were more or less oblivious to the social nature of science.[3] Carnap always held conventionalist, and in fact holist, views, his sociological orientation however seems to have been but a mere episode.[4] And while there can be no doubt about Neurath's permanently sociological interest in science, even he could occasionally slide into a terminology suggesting the image of an *individual* scientist who cannot but yield to the essentially *extrinsic* pressures of established science.[5] Notwithstanding passages like

we have to make do with words and concepts that we find when our reflections begin. Indeed all changes of concepts and names again require the help of concepts, names, definitions and connections that determine our thinking,[6]

or "even man conceived in isolation makes use of the 'intersensual' and 'intersubjective' language",[7] Neurath's formulations, generally speaking, do not seem to have a ring as radically sociological as those of Kuhn, whose position was well summarized in the formula that it is *groups*, not individuals, who should be regarded as the units which produce scientific knowledge.[8] Moreover, Neurath's views seem almost all of them open to the criticism that Kuhn's forerunner Ludwik Fleck directed against the conventionalist positions of his time. Conventionalism, wrote Fleck, fails to do justice to the fact that

the research worker has no consciousness of choice; on the contrary, the choice is imposed on him directly and in a binding manner, following from his mood of thought, from the set of his mental readinesses, from his thinking practices – in short from what we call the *thought-style* [*Denkstil*]. – The *thought-style* thus understood is the result of the theoretical and practical education of the given individual; in passing from teacher to pupil, it is a certain traditional value which is subjected to a specific historical development and specific sociological laws.[9]

The underground current running from the Vienna Circle to Kuhn cannot, then, account for certain basic tenets of the latter. Certainly there is an Austrian connection in the work of Kuhn. But this connection has to be seen in a broader perspective in at least two respects. First, it has to be seen against the background of a wider sociological turn within the philosophy of science, which ought itself to be seen as part of the overall movement, both within and without the Austrian context, towards a sociological theory of knowledge. Secondly, it has to be pointed out that the Austrian context itself encompasses more than what would normally be associated with the name of Neurath. Still, Neurath's role certainly was, in the last analysis, a symptomatic one; the following argument is therefore intended as a development, rather than as a critique, of Haller's thesis.

2. By a sociological theory of knowledge I mean something different from 'sociology of knowledge' or '*Wissenssoziologie*' as normally understood. The latter, associated mainly with names like Marx, Simmel, Lukács, Mannheim and Scheler, is primarily concerned "with the relations between knowledge and other existential factors in the society or culture".[10] As Merton writes, the sociology of knowledge "takes on pertinence under a definite complex of social and cultural conditions", namely under conditions where "conflicting perspectives and interpretations within the same society" lead to an "active and reciprocal *distrust* between groups".[11] Within a context of distrust and estrangement, the inquiry into the meaning and validity of ideas is superseded by an entirely different inquiry: by one into the motives and functions of the ideas in question. In particular, the issue of interests and ideologies plays a central

role in the sociology of knowledge. The approach I call a sociological theory of knowledge,[12] by contrast, does not deal with motives of, or interests behind, ideas; it deals with their social constitution, and is committed to the fundamental hypothesis that the ideas of the individual have no being, or at least no coherence, independent of such a social constitution.

The first one to pursue systematically a sociological theory of knowledge was Durkheim.[13] His work was, to be sure, preceded by that of the Austrian Ludwig Gumplowicz – author of the first book in German to carry the term "sociology" on its title page.[14] But Gumplowicz was primarily an early exponent of *Wissenssoziologie* rather than of a sociological theory of knowledge. He stressed the pre-determined character of individual thinking, socially inherited prejudices, "interests", but did not, apart from a few programmatic statements, attempt an epistemological re-interpretation of the concept of cogitation. Thus on the one hand he tells us that "the greatest error of individualistic psychology is the assumption that it is *man* who thinks. ... what thinks, in man, is not at all he, but his social community...".[15] On the other hand, however, he allows for the possibility of someone "emancipating" himself – albeit not before mature age, and at best to a very limited extent – from the "imprint of the mental life of *long forgotten generations*", from "thousand year old prejudices and biases".[16] And in the last analysis the social embeddedness of reason, for Gumplowicz, means merely the influence of social interests upon the mind of the individual. The thesis that social life generates thoughts and conceptions can, he says, be made more precise:

It is economic position which serves as immediate cause of the actions of the individual, presses him into adopting a certain mode of life and awakens the thoughts and conceptions which are bound up therewith.[17]

If, for Gumplowicz, the relation between society and the individual is really one of epistemological oppression, Durkheim conceives society as the very basis without which individual reason would not have been possible at all. "Nous parlons une langue", he writes,

que nous n'avons pas fait; nous nous servons d'instruments que nous n'avons pas inventés; nous invoquons des droits que nous n'avons pas institués; un trésor de connaissances est transmis à chaque génération qu'elle n'a pas elle-même amassé, etc. C'est à la société que nous devons ces biens variés de la civilisation... l'homme n'est un homme que parce qu'il est civilisé.[18]

Man is what he is through the concepts he possesses;[19] but these concepts he could not possess did he not live in society. "Qu'on essaie, par example, de se représenter", Durkheim writes,

ce que serait la notion du temps, abstraction faite des procédés par lesquels nous le divisions, le mesurons, l'exprimons au moyen de signes objectifs, un temps qui ne serait pas une succession d'années, de mois, de semaines, de jours, d'heures! ... Ce n'est pas *mon temps* qui est ainsi organisé; c'est le temps tel qu'il est objectivement pensé par tous les hommes d'une même civilisation.[20]

Conversely, any society the members of which did not share a uniform set of general concepts or "categories" would necessarily dissolve. That is why

la société ne peut-elle abandonner les catégories au libre arbitre des particuliers sans s'abandonner elle-même. Pour pouvoir vivre, elle n'a pas seulement besoin d'un suffisant conformisme moral; il y a un minimum de conformisme logique dont elle ne peut davantage se passer.[21]

Durkheim's early term *conscience collective* refers to beliefs and sentiments that are on the one hand moral and religious and on the other cognitive. Later, Durkheim introduced the term *représentations collectives* as a means of drawing a distinction between the cognitive and the moral; but faced with what he regarded as a disintegrating society, he continued to emphasize the close connection between the two. Just like moral standards, so the standards of truth and falsehood derive their binding force from the fact that they are rooted in society. As Durkheim saw, the proper carrier of concepts and ideas in the life of man is the collective. And all collective ideas have, in virtue of this origin, a quite special dignity.[22]

Three major figures can be said to have worked out, in its essentials, this sociological approach to the theory of knowledge: Durkheim's student, the sociologist Maurice Halbwachs; the Galician physician and microbiologist Ludwik Fleck, to whom we have referred already in our introduction; and the philosopher Ludwig Wittgenstein. The contribution of Halbwachs is a direct continuation of that of Durkheim. Where Durkheim relativized logical validity by conceiving it in terms of social acceptance, Halbwachs shows how social acceptance can give rise to universal standards of rationality. The notion which Halbwachs, in a work published in 1925, subjects to a radical re-interpretation is that of memory. Memory, for Halbwachs, is a function of social environment; it is invariably *collective memory*. That an individual remembers means that he moves in a frame of reference common to all the members of his —

wider or narrower – social group.[23] As Halbwachs writes:

la société ne peut vivre que si, entre les individus et les groupes qui la composent, il existe une suffisante unité de vues. ... C'est pourquoi la société tend à écarter de sa mémoire tout ce qui pourrait séparer les individus, éloigner les groupes les uns des autres, et qu'à chaque époque elle remanie ses souvenirs de manière à les mettre en accord avec les conditions variables de son équilibre.[24]

Present experience has a formative effect on the past, while conversely the past serves as the framework for present experience. Memories and new impressions will form a coherent whole, discoveries of fact are but traditions of most recent date.[25] "Ainsi s'explique que puissent s'accorder les traditions et les ideés actuelles; c'est qu'en réalité les idées actuelles sont aussi des traditions".[26] Thus ideas possess merely relative validity insofar as they are held by particular groups only. To the extent however that these groups merge into broader ones, their ideas – their transformed memories – gain broader validity. As Halbwachs puts it: "La raison s'oppose à la tradition comme une société plus étendue à une société plus étroite".[27] Rationality accordingly is not just the label we might attach to any coherent idiom of thinking and acting. Rationality is defined by the direction the amalgamation of societies actually takes; and the notion of universal reason goes hand in hand with the notion of a universal history. Durkheim, too, had made this point;[28] a rather more satisfactory elaboration of the idea seems to be possible, however, when supplemented by Halbwachs' conceptual synthesis of reason and tradition.

Ludwik Fleck's specific contribution to a sociological theory of knowledge is the argument put forward in his *Entstehung und Entwicklung einer wissenschaftlichen Tatsache: Einführung in die Lehre vom Denkstil und Denkkollektiv* (1935), to the effect that science is an *essentially* collective enterprise, a mode of investigation that would be unavailable *in principle* to an individual not initiated into the community scientific. Fleck's point of attack is the concept of a "fact". As an experienced medical scientist, he does not yield to the epistemological illusion suggested by the age-old facts of everyday life or of classical physics. He realizes that observations, tests, the diagnosis of "syndromes" etc. are not directed at something simply "given".[29] They are performed within a context of specific expectations, theoretical and practical dispositions. Fleck rejects the notion of an *experimentum crucis*, but on grounds different from those of the conventionalists. For them, scientific theories can in principle be modified freely to accommodate any experimental observation; for Fleck, in contrast, a single experiment has no practical weight whatsoever when set against the texture of over-all experience. Whereas "an experiment can be interpreted in terms of a simple question and answer, experience must be understood as a complex result of intellectual training", a training in the course of which the scientist acquires specific "physical and psychological skills", and indeed a specific, rigidly determined way of pursuing investigations in the given field.[30] As Fleck puts it, Mach and the conventionalists "pay far too little, if any, attention to the cultural-historical dependence of [the] alleged epistemological choice – the alleged convention".[31] The mind of the scientist is essentially moulded by the "thought style" into which he has been inculcated, "a thought style which determines the formulation of any concept".[32] The notion of a thought style inevitably leads to that of a "thought collective".[33] A thought collective, for Fleck, is "*a community of persons mutually exchanging ideas or maintaining intellectual interaction*". Such communities are the agents or carriers for "the historical development of any field of thought, as well as for the given stock of knowledge and level of culture".[34]

Cognition is the most socially-conditioned activity of man, and knowledge is the paramount social formation [*Gebilde*]. The very structure of language presents a compelling philosophy characteristic of the relevant community. ... the entire fund of knowledge as well as the intellectual interaction within the collective take part in every single act of cognition, which is indeed fundamentally impossible without them. ... those who consider social dependence a necessary evil and an unfortunate human inadequacy which ought to be overcome fail to realize that without social conditioning no cognition is even possible. Indeed the very word "cognition" acquires meaning only in connection with a thought collective.[35]

And Fleck's investigations into the nature of thought collectives do not stop at an abstract theoretical level. He provides concrete, illuminating sociological analyses of their structures, emphasizing in particular that within the collective quite different functions will be fulfilled by the esoteric, elite core on the one hand, and by the exoteric circle surrounding it on the other.[36] Specific influences are at work not only from the core to the periphery, but also in the opposite direction. Thus for instance the ideals of *certainty* and *simplicity* come into being not at the frontiers of research, but in the domain of popular knowledge – and the expert will merely pay lip-service to them as articles of faith.[37]

Halbwachs and Fleck re-interpreted epistemology from an external, namely from a sociological point of view; the Wittgensteinian revolution proceeds from within, dissolving the theory of knowledge into sociology by a chain of arguments which are themselves epistemological. The starting point of these arguments, elaborated by Wittgenstein from about 1930 onwards, is a criticism of the so-called name-relation theory of meaning. By showing that the meaning of a word is constituted by its *use* in language (and words are not always used as *names*) Wittgenstein can on the one hand reject mentalism (since the meaningful use of language need not be bound up with definite entities – ideas, pictures, images – in the mind of the speaker). And on the other hand he can firmly connect individual reasoning to *public criteria*, above all to those of language use. Of these criteria, however, he demonstrates that they cannot be construed as explicit rules, for then the issue of applying them would involve one in an infinite regress. Ordered use thus points to rules *embedded in behaviour*, to *uniform ways of acting*,[38] to a consensus not only of opinions but of forms of life.[39] As Wittgenstein wrote in one of his last remarks: "We do not learn the practice of making empirical judgments by learning rules: we are taught *judgments* and their connexion with other judgments. A *totality* of judgments is made plausible to us".[40]

If we now compare the respective positions of Durkheim, Halbwachs, Fleck and Wittgenstein in greater detail, then there are many points of striking similarity which meet the eye. Thus e.g. Durkheim, Halbwachs and Wittgenstein all allot a special role to something like *a priori* certainties. Durkheim speaks of "necessary ideas" which we accept "without any proof" since behind them stands the authority of society.[41] Halbwachs speaks of the "immovable frames of reference" in which memories move and which are given to us from without.[42] Wittgenstein speaks of "hinges" that "must stay put", of facts which are "fused into the foundation of our language-game".[43] Such certainties are the "common place, where every one meets",[44] the "common way", which all must take,[45] the direction in which "one as it were hurries to a meeting with the others".[46] The basic relations pertaining to categories possess, for Durkheim, "a specific kind of moral necessity";[47] Wittgenstein writes about the laws of logic as constraining us, like other laws in society.[48] Halbwachs, Fleck and Wittgenstein all emphasize the epistemological role our *schooling* plays. Halbwachs speaks about the "influence of education, of teaching", transmitting "uninterruptedly and from early on" the intellectual habits of society;[49] Fleck ridicules the cliché that "empirical experience" is the origin of "man's" knowledge, when actually "for a very long time, the source of almost the entire knowledge of every

man was, in Europe, simply book and school";[50] Wittgenstein contrasts
"what we learnt at school" with the "transcendent certainty" of the
philosophers, and writes:

I learned an enormous amount and accepted it on human authority, and then I found
some things confirmed or disconfirmed by my own experience. ... For how can a
child immediately doubt what is taught? That could mean only that he was incapable
of learning certain language games. ... "We are quite sure of it" does not mean just
that every single person is certain of it, but that we belong to a community which is
bound together by science and education.[51]

Man is "trained" to *understand* what he sees and feels,[52] "traditional
patterns of training" are involved in scientific experience,[53] teaching a
child its first language takes place not by explanation but by drill.[54] The
possibility of individual creativity is explained both by Halbwachs and by
Fleck by reference to the individual's being a member of more than one
thought-community at the same time,[55] an idea paralleled by Wittgenstein's notion of a "superimposition of several languages".[56] Both
Halbwachs and Wittgenstein deny that mental processes are in any way
obscure or opaque. Memories, Halbwachs points out, "have nothing
mysterious about them. If we ask where they are, where they are stored,
in my head, or in some corner in my mind to which I alone would have
access, then there is nothing here to be found".[57] Thinking, writes
Wittgenstein, is not a "secret and blurred process", of which we see only
"indications" in language, for in "*thought* there is nothing private".[58]
Thinking is a *transparent* phenomenon, since its typical vehicle, language, is public.[59] The assumption of an "inner language" will not do,
since, as Halbwachs puts it, such a language, lacking the "control of
society", would display no order.[60] Wittgenstein's arguments against the
possibility of a private language start from a similar observation: "to *think*
one is obeying a rule is not to obey a rule. Hence it is not possible to obey
a rule 'privately': otherwise thinking one was obeying a rule would be the
same thing as obeying it".[61] As Halbwachs formulates it, we are able to
remember only if "the reliability of our memory" is "guaranteed" by our
bond with "a group of people".[62] Mental *images*, as Halbwachs points
out, anticipating Wittgenstein, will not do as standards against which to
compare the use of words: they are but "rough materials" that can be
"freely combined";[63] indeed words can better describe images than can
images represent words.[64] And there is an important similarity in the way
both Halbwachs and Wittgenstein take individual idiosyncrasies to be
largely dependent on the nature of the given social environment.

Halbwachs imagines a society "in which the meaning of words was indeterminate and subject to constant change": in such a society someone with speech abilities that are normal in our sense would soon find himself in the situation of an aphasiac.[65] And Wittgenstein asks: "What would a society all of deaf men be like? Or a society of the 'feeble-minded'? *An important question!* What, then, would a society be like that never played many of our customary language-games?"[66] If, in a culture like ours, remarks Wittgenstein, a child does not respond to the suggestive gestures of his teacher and does not make the transition from, say, "20" to "21", then "it is separated from the others and treated as a lunatic".[67] One could however imagine a tribe in the life of which a certain numeral played a specific role, that of "an unsurmountable limit",[68] and then of course the above-mentioned child might count as perfectly normal. Even a question such as the correctness or incorrectness of an empirical observation is to a great extent dependent on cultural factors. Both Fleck and Wittgenstein point out, for example, that there is no such thing as an "exact" or "inexact" measurement in the absolute sense. Given a task of a certain complexity, how many single measurements are to be carried out to approach ideal exactness, and what scatter is admissible? As Fleck insists, we make the decision "on the basis of custom, the entire store of individual and collective knowledge".[69] Or as Wittgenstein puts it: "No *single* ideal of exactness has been layed down: we do not know what we should be supposed to imagine under this head".[70] Or again, it depends on the dominant thought-style which *Gestalten* or patterns will be picked out in the course of any given observation. As Fleck writes: "A *Gestalt* is constructed not from 'objective physical elements', but from cultural and historical themes".[71] Wittgenstein, too, emphasizes the sociological element involved in seeing something *as* something – in seeing it as this or that *Gestalt*: "Now does one teach a child (say in arithmetic)", he asks,

"Now take *these* things together!" or "Now *these* go together"? Clearly "taking together" and "going together" must originally have had another meaning for him than that of *seeing* in this way or that. – And this is a remark about concepts, not about teaching methods.[72]

And with respect to scientific advances, both Fleck and Wittgenstein believe that the image of the scientist as a *discoverer* (of new "facts", awaiting discovery) is grossly misleading. "The textbook", writes Fleck, "changes the subjective judgment of the author into a proven fact. ... Social distance transforms the author from a creator to a discoverer".[73] In a similar vein does Wittgenstein say, in speaking of mathematics, that "the mathematician is an inventor, not a discoverer".[74]

Given such parallels, now, the question naturally arises as to what sort of influences there might obtain between the figures mentioned. In the case of Halbwachs it is of course Durkheim's work against the background of which his own contributions have to be interpreted. An indirect influence by Durkheim is conspicuous in the case of Fleck, and can be assumed, with some plausibility, in the case of Wittgenstein. Fleck encountered the notion of collective representations in the German edition of *Les fonctions mentales dans les sociétés inférieures* by Durkheim's student Lévy-Bruhl, to which edition Wilhelm Jerusalem provided a preface.[75] About Gumplowicz – who was at some stage himself influenced by Durkheim[76] – Fleck learned by reading Jerusalem's "Die soziale Bedingtheit des Denkens und der Denkformen" in the collection edited by Max Scheler, *Versuche zu einer Soziologie des Wissens.*[77] Jerusalem's term "soziale Verdichtung" Fleck finds useful,[78] but is otherwise critical of both him and Lévy-Bruhl because they show an all too great respect for "scientific facts".[79] Wittgenstein's manuscripts contain no reference to Durkheim. However, he might have been exposed to a Durkheimian influence through the classical scholar and philosopher of language Nikolai Bakhtin, to whom he was very close, and who studied at the Sorbonne before settling in Cambridge in 1932.[80] It was indeed in the period between 1931 and 1936 that there occurred a shift in Wittgenstein's position from the initial anti-mentalism of his so-called later philosophy to a more markedly *sociological* theory of knowledge. Through Bakhtin Wittgenstein might even have heard about Halbwachs' work: but this is merely speculation. He was certainly not aware of Fleck. Nor is there the slightest reason to suppose that Fleck knew about Halbwachs' book when writing his own, or that Halbwachs had cognisance of the fact that Fleck's work had appeared in 1935. But a few years later they did, for some months, come into close proximity. Both were deported to Buchenwald. Halbwachs died there in March 1945; Fleck survived, returned to Poland, and from there he was permitted, in 1957, to emigrate to Israel.

3. In fact not only Halbwachs and Fleck, but all the other figures who play a role in the development of *Wissenssoziologie* and of a sociological theory of knowledge – Marx, Gumplowicz, Durkheim, Jerusalem, Simmel, Scheler, Lukács, Mannheim, Neurath, Wittgenstein, Merton,

Kuhn – are of Jewish origin. And it is important to realize that with regard to most of them one can point to essentially Jewish experiences which were formative in their development and sometimes came to haunt them. Durkheim's father was a rabbi, as his grandfather and his great-grandfather had been before him; he grew up "within the confines of a close-knit, orthodox and traditional Jewish family, part of the long-established Jewish community of Alsace-Lorraine".[81] Durkheim, too, was destined for the rabbinate, and for a time he studied at a rabbinical school. The family atmosphere was one where, as Lukes quotes Davy, "the observance of the law was precept and example", a background instilling in Durkheim an aversion to "the life of the individual" not positively rooted "within the framework of the group". "From the time of his childhood", writes Lukes, Durkheim "retained an exacting sense of duty... he would never experience pleasure without a sense of remorse". On the other hand he undoubtedly experienced "what he later described as 'that tempering of character, that heightening of life which a strongly cohesive group communicates to its members'".[82] Gumplowicz's attitude to Jewishness was quite different. He, too, descended from a family of rabbis, but he expressed a strong distaste for the persistence of Jewish separateness, for the Jews' clinging to an "obsolate culture".[83] He himself converted to Protestantism in the early 1870s.[84] Of Otto Neurath, by contrast, it is remarked that "in spite of all his looking towards the future" there was in him a "full awareness of the past, of a strong tradition in the family that went back for generations".[85] That tradition was Gentile on the side of his mother, and Jewish on the side of his father, Wilhelm Neurath, Dr.phil., whose parents were, as the latter puts it in his autobiography, "poor, very religious people of strictest morals", so much so that they feared "worldly knowledge" might shake his belief. Wilhelm Neurath hungered for that knowledge and went out, on his own, to learn. He became a materialist and an atheist, and eventually a scholar of renown. But then there occurred "emotional and physical upsets" which led him to form a "pantheistic-theistic conception of the world".[86] and it was in this peculiar atmosphere of learning, science, and latent mysticism that Otto Neurath grew up. His third wife Marie Neurath characterized him, as well as his first wife Anna ("a Polish Jew, very shy and lonely when they first met"), as "'outsiders' of society".[87] Another outsider, a "stranger", was Georg Simmel,[88] born of Jewish parents, but baptized in, and brought up according to the norms of, the Lutheran church. However, he was never fully accepted by his Gentile academic peers.[89] Nor did

Simmel's student Max Scheler really belong to any community. The young Scheler, an ardent Jewish believer, became gradually attached to Catholicism and was admitted to the Church at the age of 25; subsequently he was tormented by doubts, but experienced something like a second conversion; a few years later he finally lost faith in the integrating force of the Catholic *Weltanschauung* and left the Church. Simmel's student Georg Lukács, growing up in a rich Jewish family in Budapest, professed no religious convictions; but he felt an acute nostalgia for the lost age of closed communities, and was feverishly searching for ideas of a future – ideas not excluding Jewish messianism – that would bring back the past.[90] And in Wittgenstein's case the issue of his own Jewishness (he was three-quarters Jewish and baptized in the Catholic Church) was a burning personal problem during those very years in the course of which his later philosophy was developed.

The question thus arises whether Jewish descent could have had something to do with the conception of a sociological theory of knowledge. Two sorts of answers suggest themselves immediately. First, Jewish emancipation liberated a vast reservoir of intellectual energies in all fields,[91] and sociology and its various subdisciplines were in this respect no exception. Secondly, sociology was, around the turn of the century and for some decades after, a newly-established discipline, offering greater than average opportunities to those who were, socially speaking, newcomers. In fact, the share of scholars of Jewish origin in the field of sociology is truly remarkable. But here a different consideration, too, already enters the picture: *viz.* the obvious insight that the continuity of Jewish history in the Christian era – with its dependence on social organization and community cohesion rather than on religious hierarchies, political structures and state power – could easily have suggested, to emancipated Jews who entered the social sciences, a specifically *sociological* point of view. And this consideration then leads to the particular question I would like to formulate here (without of course attempting to answer it in any detail): does it make sense to speak of some specific trait or traits of Jewish forms of life or modes of thinking which would or could give rise to the perspective characteristic of a sociological theory of knowledge – the perspective that individual reason is inconceivable outside the framework of interpersonal relations? Durkheim himself characterized Judaism as consisting, essentially, "comme toutes les religions inférieures ... en un corps de pratiques qui réglementent minutieusement tous les details de l'existence et ne laissent que peu de place au jugement individuel", and pointed, with respect to Jewish communities, to the "nécessité de lutter contre une animosité générale, l'impossibilité même de communiquer librement avec le reste de la population".[92] Within the community, by contrast, there is a steady and lively communication, even on a fairly learned level, as to the interpretation and application of handed-down laws in situations new, unclear, or contradictory. Thus Leo Baeck could speak of a "community of thinkers",[93] with the emphasis however on the word "community":[94] ideas are conceived as intersubjective rather than subjective, intellectual education as a "building up", rather than as a liberation of some innermost tendencies, religious commitment as a question of obeying norms rather than as experiencing the "inward feeling".[95] Not only the relation between God and man has a dialogic structure,[96] but human individuality in itself too: the "I" is not conceivable without a "Thou", without a fellow-man.[97] The existence and continuity of Jewish peoplehood is sustained, its borders are delimited,[98] by a process of religious and intellectual communication.

Against the background here sketched the emancipated Jewish intellectual can entertain, abstractly speaking, a negative or a positive attitude with respect to the place of individual reason within a system of collective beliefs. The former attitude can, again, take the form of what one might call two different moods, that of liberatedness and that of alienation. For the liberated mood, nothing befits individual reason better than to wage a constant battle against the authority of received views, to critically replace old ideas by new ones. This is the mood of a Karl Popper,[99] or indeed of any philosopher of Jewish origin who chooses to ignore the social context of cognition – one thinks of Edmund Husserl, or even of Alfred Schütz (whose "phenomenology of the social world" is built up by analyses of individual intentional acts of the Husserlian-Weberian type[100]), and indeed even of a Wilhelm Jerusalem, notwithstanding his half-hearted concessions to a sociology of knowledge. For the alienated mood, the mood of the "stranger", in contrast, the individual's thinking is in fact, and hopelessly, determined by what its group believes and *is*; one group's truths might be unmasked as ideologies by members of another group, but the latter, too, is held captive by ideologies of its own; only someone not belonging to any community at all – Mannheim's "free-floating intellectual" – is able to strive towards a non-ideological view of the world. This is the mood of *Wissenssoziologie* – from which only a messianistic twist will offer redemption. The positive attitude, now, is the one that can give rise to a secular interpretation of what is felt to be the truth in Judaism's epistemological message: collective reason is constitutive of individual minds, but has no reality independent of the latter.[101] Let us now, by way of conclusion, connect these last considerations with our initial theme. With respect to Austrian philosophy in general the Jewish contribution has certainly been taken note of in modern historical research,[102] even if this aspect of the story has more recently somewhat receded into the background. However, if what I have suggested has any air of plausibility, some more straightforward implications for our understanding of the development of Austrian philosophy will have to be faced. In particular, the usual bifurcation of the pre-history of the Vienna Circle into a Machian and a Brentanian line[103] will have to be supplemented by a division of a different kind: by the triple distinction of Catholic, Protestant, and Jewish roots in the make-up of the philosophy of science as we know it.

CHAPTER 4

SOME MARXIAN THEMES IN THE
AGE OF INFORMATION

The radical conceptual and social changes related to recent developments in technology – aptly summed up under the term "computer revolution" – give new interest to certain basic tenets of Marx. But these tenets are at the same time moved into a perspective hardly envisaged by Marx himself.

There is, to begin with, one fundamental Marxian thesis which can definitely be said to have gained in plausibility. This is the thesis of *technological determinism* – relativized already by Engels,[1] contested by scholars of such different persuasions as Max Weber[2] or Georg Lukács,[3] but no doubt very strongly adhered to by Marx himself. Recall the well-known passage in *The Poverty of Philosophy*:

In acquiring new productive forces men change their mode of production; and in changing their mode of production, in changing the way of earning their living, they change all their social relations. The hand-mill gives you society with the feudal lord: the steam-mill, society with the industrial capitalist.[4]

And recall also the famous preface to the *Critique of Political Economy*, where Marx affirms that the way men produce their means of subsistence conditions their whole social, political and intellectual life.[5] That developments in microelectronics have deep implications for society and politics in the United States, in Western Europe, and in the Far East, is by now of course obvious. But it is obvious, too, that these implications are not independent of the social and political frameworks within which they emerge. There is one part of the world, however, namely Eastern Europe and the Soviet Union, with regard to which the political and social effects of the new technology are absolutely determining in the sense that they led to changes to which the dominant political forces have been rigidly opposed; to changes for which the inherited and existing political cultures in the countries in question – with no democratic traditions whatsoever in the Soviet Union, and with a bare minimum of such traditions in Eastern Europe – did not provide a framework, and did not herald any promise.

Has not recent history amply demonstrated that the dream of drawing closer to the West – characteristically not indulged in by Russians – could at any time be turned into a nightmare in Eastern Europe? But then came the *chip*, and a very different picture emerged. The centralized economies of the so-called socialist countries proved to be unable to keep up with Western developments in microelectronics. As a consequence, the faith in the continuing military supremacy of the Soviet Union over the West wavered. Eastern European products became, for reasons of quality and price, increasingly difficult to export. Attempts at a decentralization of economy with no democratization of the political system failed. Liberalization at home and a new *détente* in foreign policy were the result. It would require a great deal of naivety not to see that in this instance, once more, a deeply Marxian idea has been confirmed: namely the idea that instead of the personal traits of political leaders forming their policies, it is, much rather, the exigencies of political realities that become reflected in the personal make-up of politicians.

Seen from a Marxist perspective, however, this state of affairs possesses a truly strange feature. For the economy providing a suitable framework for technological progress thus turns out to be not that of central planning, but that of the free market. Indeed the situation, as in particular the Japanese experience shows, is even more peculiar: the presence of some old-fashioned traditions in the texture of a liberal society does *not* seem to be an obstacle to the development of successful free enterprises, and, by implication, to advances in technology. And the joint values of the free market on the one hand and of traditionalism on the other add up to just about everything Marx *detested* – and to just about everything Marx's arch-liberal adversary F. A. von Hayek stands for.[6]

Some central Marxian themes are, to be sure, only seemingly affected by the emergence of the new technologies. Thus for instance the labour theory of value, with all its implications, strikes one today, at a time when knowledge has become the supreme commodity, as utterly implausible; but this theory had never been a logically acceptable one, and was, precisely with reference to the effects of science, withdrawn by Marx himself. The plausibility of the labour theory of value has radically decreased because the element of labour *time* has lost its relevance as a source of added value. The substance of value, Marx held, is labour; its *measure* is time.[7] As exchange values, all commodities are but certain amounts of congealed labour time.[8] Today however it is obviously *knowledge*,[9] not labour time, that is primarily embodied in the added value of any commodity.[10] Incidentally, this was already the case, even if to a lesser degree, in Marx's days – a state of affairs he fully realized,[11] but to which he gave a strangely twisted interpretation. As Marx saw it, the labour theory of value would cease to be valid once the capitalist mode of production had been superseded; and the signs that the theory *was* in fact becoming increasingly implausible he understood as heralding the imminent doom of capitalism. In this sense the labour theory of value was not susceptible to scientific refutation;[12] only the historical deed of establishing communism could prove it false – by rendering it obsolete. And it is indeed a piece of irony that the attempt at that historical deed led to an entirely different result: to the perfect realization of the labour theory of value, in the form of the Soviet *labour camp*. There, certainly, all work was reduced to uniform, simple labour, measurable in units of men and time.

Then there are other Marxian convictions, for instance some of those having to do with the idea of the concentration of capital, which, for a long time, seemed convincing, but appear antiquated in the light of today's high-tech economy, in particular in the light of developments in software engineering. Although huge enterprises obviously do play an important, sometimes dominant, role in the electronics industry, and although with the increasing tendency of programs to be written by large teams the costs of software production are, on the whole, rising, it is still the case that small firms in these areas continue to have relatively good chances of success, and that the software industry still offers entrepreneurial possibilities for programmers with practically no capital to invest. The success stories of bright teenager "hackers" – virtuoso programmers – in the United States and in Western Europe are a familiar theme. And it is significant that software development is practically the only economic domain where a backward country like Hungary, with no funds to mobilize, seems to be able to co-operate effectively, and in places to compete, with Western firms.

On the other hand the romantic-eschatological aspect in Marx – notoriously a cause of embarrassment to bureaucratic Marxism – is today acquiring an air of reasonableness. Certainly the overthrow of liberal institutions and the elimination of free competition by a revolutionary proletariat no longer has, for those who tried the Marxist experiment, the ring of promise it apparently possessed earlier. Yet the emerging technologies for handling information do indeed seem to herald a new age of community, of the *vergesellschafteter Mensch* – of participatory democracy, of a new craftsmanship, of non-alienated cultural patterns.

Discussing the historical role of a mode of production based on exchange
values, Marx writes:

The universal nature of this production with its generality creates an estrangement of
the individual from himself and others, but also for the first time the general and
universal nature of his relationships and capacities. At early stages of development the
single individual appears to be more complete, since he has not yet elaborated the
wealth of his relationships, and has not yet established them as powers and
autonomous social relationships, that are opposed to himself. It is as ridiculous to
wish to return to that primitive abundance as it is to believe in the necessity of its
complete depletion.[13]

Clearly, Marx's vision of a non-alienated past does indeed play a part in
his dream of a non-alienated future.[14] And when depicting that future he
strikes a utopian, almost millenial note. The historical mission of
capitalism is fulfilled and a new age begins when the productive forces of
labour have reached a stage at which general affluence is maintained by a
minimum of labour essentially *scientific*, indeed by an activity which is
really the free development of rich personalities.[15]

Now the specialist whose work is most deeply embedded in, and is
perhaps most revealing of, the age of information, is the professional
programmer. Thus in assessing the claim that the tendencies emerging in
this age in a sense vindicate the Marxian utopia, it seems reasonable to
begin by analyzing the way the members of this profession labour and
live. First impressions are, certainly, discouraging. As Sherry Turkle
writes:

In the course of the last decade programmers have watched their opportunities to
exercise their expertise in a spontaneous way being taken away. Those who are old
enough remember the time when things were different as a kind of golden age, an age
when a programmer was a skilled artisan who was given a problem and asked to
conceive of and craft a solution. ... Today, programs are written on a kind of
assembly line. ... Thus programmers are particularly sensitive to the fragmentation of
knowledge and the lack of a feeling of wholeness in work to which so many of us are
subject.[16]

But this is not the only possible perspective. As David Bolter puts it:

The computer shows that even teamwork need not thoroughly subsume and
homogenize the special contribution of each member. The best organization for many
computer projects is modular: each member of the group is given a separate part of
the larger program or machine design. This is not the stultifying specialization of the
assembly line, where one worker performs one operation repeatedly for hours.
Instead, each module may be a self-contained program or portion of hardware, with
challenges and difficulties all its own.[17]

Another way to point to the non-alienating aspects of the computer is to highlight its *tool-like* character. "The computer is", writes Bolter,

in some ways a grand machine in the Western mechanical-dynamic tradition and in other ways a tool-in-hand from the ancient craft tradition. The best way to encourage the humane use of computers is to emphasize, where possible, the second heritage over the first, the tool over the machine. – A machine is characterized by sustained, autonomous action. It is set up by human hands and then is more or less set loose from human control. ... A tool, unlike a machine, is not self-sufficient or autonomous in action. It requires the skill of a craftsman...[18]

Now Turkle, too, exploits – with reference to Marx[19] – the tool–machine distinction. "Tools are extensions of their users; machines", she writes, "impose their own rhythm, their rules, on the people who work with them, to the point where it is no longer clear who or what is being used".[20] At work – in the office, at the lab – the computer has become a machine; but *at home* – this is the aspect Turkle stresses – it can play the compensatory role of a tool. When people in the electronics industry, or professional programmers, speak of the way they approach problems on their home computers – in their free time, as a hobby – they convey "a sense of power" that comes from "having full knowledge of the system", from working in a "safe environment" of their "own creation".[21] Building up from machine code to finished project, becoming directly involved, as it were, with the workings of the CPU – the central processing unit – itself, turns the computer virtually into a physical tool.

The CPU's primary activity is moving something that is conceptually almost a physical object (a byte of information) in and out of something (a register) that is conceptually almost a physical place. The metaphor is spatial, concrete. One can imagine finding the bytes, examining them, doing something very simple to them, and passing them on. ... People are able to identify physically with what is happening inside the machine. It makes the machine feel like a part of oneself. It encourages appropriation of the machine as a tool in Marx's sense – as an extension of the user.[22]

The idea that it is the worker's *free time* which constitutes the true domain of non-alienated activity is of course again a very Marxian one, one belonging to the less romantic layers of his thinking.[23] But it appears that it is precisely the romantic-utopian vision of the *Grundrisse* which today is becoming increasingly plausible. The emergence of the *homo otiosus*, the human being enjoying the leisure of his free time, will, so it seems, coincide with developments which lead to a blurring of the boundaries between working hours and the hours spent off work.[24] The main new element here is the possibility of working at home, the "return to cottage industry on a new, higher, electronic basis, and with it a new emphasis on the home as the center of society".[25] Economically this might imply, as Toffler puts it, that

if individuals came to own their own electronic terminals and equipment, ... they would become, in effect, independent entrepreneurs rather than classical employees – meaning, as it were, increased ownership of the "means of production" by the worker.[26]

The possible sociological implications are no less significant:

If employees can perform some or all of their work tasks at home, they do not have to move every time they change jobs, as many are compelled to do today. They can simply plug into a different computer. – This implies less forced mobility, less stress on the individual, fewer transient human relationships, and greater participation in community life. ... The electronic cottage could mean more of what sociologists, with their love of German jargon, call *gemeinschaft*.[27]

It is not by their indirect effects on the local level however, but by their direct impact on a nationwide or even a global one, that computer networks contribute most significantly to the forming and maintaining of new communities. Discussing the introduction of personal computers in the late 1970s, Sherry Turkle points out that

they came on the scene at a time of dashed hopes for making politics open and participatory. Personal computers were small, individually owned, and when linked through networks over telephone lines they could be used to bring people together. ... The computer clubs that sprang up all over the country were imbued with excitement not only about the computers themselves, but about new kinds of social relationships people believed would follow in their wake. ... Personal computers became symbols of hope for a new populism in which citizens would band together to run information resources and local government.[28]

Such hopes might have been premature at the time; but they were certainly not delusive in principle. For computer networks can in fact become instrumental in overcoming the information gap separating the knowledge any individual has from the knowledge society at large possesses.[29] This information gap is, indeed, a real source of alienation in the modern world. In closed, pre-literate societies the knowledge conveyed by oral traditions was partly knowledge in the common realm; partly knowledge available to the initiated only, but to them in fact directly accessible. With the rise of literacy – a fundamental change in the technology of communication, information storage and retrieval[30] – knowledge became embodied in written texts. And the first early libraries contained in principle *all* there was to read: the information they provided was a global one. Even a hundred years ago it was still possible for someone to assume that he was acquainted with all the essential documents that were of importance for his private and professional life. Contemporary man however has lost control over his informational environment.[31] Computer networks – representing a second fundamental change in the technology of communication – are a possible means to regain that control.

CHAPTER 5

TRADITION AND PRACTICAL KNOWLEDGE

The first task of this chapter is to indicate how the topic of practical knowledge might involve, or why it should involve, an analysis of the notion of tradition. Such an indication is in fact not difficult to give. After all, both practical knowledge and knowledge embedded in tradition are kinds of knowledge that seem to lie outside the domain of reflection or reasoning; both presuppose an epistemological subject whose activity encompasses more than the life of pure cognition – a subject to whose make-up traits other than mental essentially belong. No wonder, then, that philosophers with an eye for the dimension of practice in knowledge will usually not fail to draw attention also to the special ways in which that dimension is transmitted: to ways of custom, to institutions of handing down, to traditions.

Thus Ryle stresses that learning *how* is different from learning *that*: the former involves, as the latter does not, inculcation,[1] i.e. persistent, impressive repetition. Thus also Michael Polanyi, after having argued that the rules of scientific discovery are but rules of art, goes on to point out that since "an art cannot be precisely defined, it can be transmitted only by examples of the practice which embodies it".[2] Science, he writes at another place, "is operated by the skill of the scientist",[3] by a skill that, again, can only be passed on by example. But to learn by example is "to submit to authority. ... By watching the master and emulating his efforts in the presence of his example, the apprentice unconsciously picks up the rules of the art, including those which are not explicitly known to the master himself. These hidden rules can be assimilated only by a person who surrenders himself to that extent uncritically to the imitation of another", by a person who will "submit to tradition".[4]

Oakeshott, too, points out that "the coherence of scientific activity" does not "lie in a body of principles or rules to be observed by the scientist, a 'scientific method'"; that coherence, he stresses, "lies nowhere but in the way the scientist goes about his investigation, in the traditions of scientific inquiry".[5] And one of the main claims of T. S. Kuhn is of course that "we have too long ignored the manner in which knowledge of nature can be tacitly embodied in whole experiences without intervening abstraction of criteria or generalizations. Those experiences are presented to us during education and professional initiation by a generation which already knows what they are exemplars of".[6] This seems also to be the idea taken up by David Bloor when he writes (referring, incidentally, to Mary Hesse's *Structure of Scientific Inference*):

Predicates are learnt on the basis of a finite number of instances. These are provided by teachers or authorities who must simultaneously inform and control the behaviour of the learner. The learner's task is to acquire a sense of the similarity between the cases to which he is exposed as instances of a given concept. His sense of similarity and difference must be matched to those of other language users. This involves grasping the *conventions* which are involved in the judgements about similarity and difference.[7]

Even Feyerabend, having, in *Science in a Free Society*, once more made his peace with Wittgenstein, writes of "standards or rules" we could not use were they not "well integrated parts of a rather complex and in places quite opaque practice or tradition".[8] As to Wittgenstein himself, one need recall only the central role his arguments played in turning into a philosophical issue the idea of knowledge embedded in, or constituted by, practice. When von Wright, interpreting Wittgenstein's *On Certainty*, coined the notion of "pre-knowledge", remarking of course that the same "is not propositional knowledge" but rather a *praxis*,[9] the profession was quick to suggest that the appropriate term here might be not "pre-" knowledge, but, rather, *practical knowledge*.[10] And I would like to underline that in those arguments of Wittgenstein in which the idea of practical knowledge essentially figures, the concept of tradition, too, inevitably crops up, expressed by the terms *Gepflogenheit, Gebrauch, Institution, Lebensform*, or *Autorität*.[11]

My point of departure is, then, roughly this: since practical knowledge encompasses, or serves as a foundation for, much of what we know, and since such knowledge appears to be tacit, non-propositional, and indeed inarticulable,[12] channels of communication other than explicit discourse have indispensable functions to fulfil. Traditions represent just such channels.

That this initial position immediately leads to a number of questions and difficulties, is clear. The first such difficulty is presented by the notion of practical knowledge itself, which seems on occasion precisely not to require any social context of transmission. Take skills, for example. Clearly skills are, or embody, practical knowledge; but not all skills presuppose a social context. Thus cycling, one of Polanyi's favourite examples,[13] involves a vast amount of tacit knowledge, in the sense that the mathematical description of what happens at every moment whilst one adjusts the curvature of one's bicycle's path in proportion to the ratio of one's imbalance over the square of one's speed is of course unknown to the cyclist, and would not help him in his performance even were it known. But I don't see what is, in principle, unarticulable about this knowledge; and I certainly cannot recall anything like a state of apprenticeship when learning to ride my first bicycle. I saw what other people were doing, but I did not learn by imitating them, I learnt by constantly falling, and then sometimes not falling, off. It seems there are *technical* skills – like cycling – and *social* skills – like speaking, or counting – and the former do not presuppose a tradition in the immediate sense in which the latter do. Or take medical diagnosis, another of Polanyi's examples. "Unless a doctor can recognize certain symptoms", writes Polanyi,

e.g. the accentuation of the second sound of the pulmonary artery, there is no use in his reading the description of syndromes of which this symptom forms part. He must personally know that symptom and he can learn this only by repeatedly being given cases for auscultation in which the symptom is authoritatively known to be present, side by side with other cases in which it is authoritatively known to be absent, until he has fully realized the difference between them and can demonstrate his knowledge practically to the satisfaction of an expert.[14]

It was similar or related observations that led Ludwik Fleck in the early thirties to his traditionalist, pre-Kuhnian theory of science. Thus in his explanations of the Wassermann reaction, Fleck points out that the "reaction occurs according to a fixed scheme, but every laboratory uses its own modified procedure, which is based upon precise quantitative calculations; nevertheless, the experienced eye or the 'serological touch'" – *das serologische Fühlen* – "is much more important than calculation".[15] The field of serology, Fleck writes, "is a little world of its own and therefore can no more be fully described in words than any other field of science".[16]

It is however a fact that important areas of medical diagnosis are today conducted by computer programs, and it would seem strange to speak of "personal knowledge" or "touch" with respect to a piece of software. Of course these programs are based on the knowledge of experienced human experts, and it is in fact quite a problem to unearth that knowledge in software-digestible form. One becomes an expert not simply by absorbing explicit knowledge of the type found in textbooks, but through experience, that is, through repeated trials, "failing, succeeding, wasting time and effort, ... getting a feel for a problem, learning when to go by the book and when to break the rules". Human experts thereby gradually absorb "a repertory of working rules of thumb, or 'heuristics', that, combined with book knowledge, make them expert practitioners".[17] This practical, heuristic knowledge, as attempts to simulate it on the machine have shown, is "hardest to get at because experts – or anyone else – rarely have the self-awareness to recognize what it is. So it must be mined out of their heads painstakingly, one jewel at a time".[18]

But now, practical knowledge as here described does not seem to possess any philosophically interesting characteristics at all, and it is quite disturbing to realize that the faculty of judgment, the ability to subsume particular instances under a given rule, or the ability to apply rules, can be imparted to a suitable machine without further ado, without extended training on the learner's side, without the full social context that seemed so essential for this kind of acquisition.

For Kant already the application of rules seemed to embody a specific philosophical problem:

If understanding in general is to be viewed as the faculty of rules, judgment will be the faculty of subsuming under rules; that is, of distinguishing whether something does or does not stand under a given rule... General logic contains, and can contain, no rules for judgment. ... If it sought to give general instructions how we are to subsume under these rules, that is, to distinguish whether something does or does not come under them, that could only be by means of another rule. This in turn, for the very reason that it is a rule, again demands guidance from judgment. And thus it appears that, though understanding is capable of being instructed, and of being equipped with rules, judgment is a peculiar talent which can be practised only, and cannot be taught. It is the specific quality of so-called mother-wit... Deficiency in judgment is just what is ordinarily called stupidity, and for such a failing there is no remedy. ... A physician, a judge, or a ruler may have at command many excellent pathological, legal, or political rules, even to the degree that he may become a profound teacher of them, and yet, none the less, may easily stumble in their application. For, although admirable in understanding, he may be wanting in natural power of judgment. He may comprehend the universal *in abstracto*, and yet not be able to distinguish whether a case *in concreto* comes under it. Or the error may be due to his not having received, through examples and actual practice, adequate training for this particular act of judgment. Such sharpening of the judgment is indeed the one great benefit of examples.[19]

Ryle, too, stresses that *stupidity* is not the same as mere ignorance, pointing out that:

The consideration of propositions is itself an operation the execution of which can be more or less intelligent, less or more stupid. But if, for any operation to be intelligently executed, a prior theoretical operation had first to be performed and performed intelligently, it would be a logical impossibility for anyone ever to break the circle.[20]

Similar infinite regress arguments play a central role in Wittgenstein's later philosophy,[21] nor are they missing from Polanyi's writings – "The application of rules must always rely ultimately on acts not determined by rule"[22] – or, for that matter, from F.A. von Hayek's:

there will always be some rules governing a mind which that mind in its then prevailing state cannot communicate, and ... if it ever were to acquire the capacity of communicating these rules, this would presuppose that it had acquired further higher rules which make the communication of the former possible but which themselves will still be incommunicable.[23]

But it is exactly this infinite regress argument, seemingly so central to all philosophizing about practical knowledge, which somehow looses its magic once the nature of knowledge built into non-human expert systems has been considered.

Or take the case of Ryle's "well-trained sailor boy", who "can both tie complex knots and discern whether someone else is tying them correctly or incorrectly, deftly or clumsily. But he is probably incapable of the difficult task of describing in words how the knots should be tied".[24] Knots are more easily tied than explained, but the boy's presumed inability to do the latter does not seem to carry a philosophical message. He might be unable to explain *anything*. Or a detailed terminology of knots could be developed, helped by which the boy would have no difficulties at all in describing and criticizing. Of course the usual way to explain tying knots is through pictures rather than through words. And here one should perhaps say that though knowledge conveyed through pictures might be non-propositional, it does not therefore necessarily follow that it is practical, i.e non-theoretical, in the sense of the present discussion.

It might be useful, at this stage, to distinguish between two positions with regard to the issue of practical knowledge. According to the first, this knowledge is a *practical abbreviation* within the texture, or flow, of knowledge as such; a device of paramount pragmatic importance perhaps, but not something the discovery of which should basically transform our epistemological convictions. According to the second position, there is a layer, or dimension, of practical knowledge which could in no sense be dissolved into knowing that. Or perhaps – and this would be a stronger version of the same position – there is a hard layer of practical knowledge which serves as the bedrock upon which *all* knowledge rests. Or indeed – to formulate a yet stronger version – all theoretical knowledge represents but an articulating, a spelling out, of a knowledge which is, in the last analysis, invariably reducible to practice. Philosophers like Wittgenstein, Oakeshott, or Kuhn, clearly hold some version of the second position; but Ryle, too, flatly states that "theorising is one practice amongst others".[25]

Now each of these positions has its counterpart within the theory of traditions. Let us distinguish between *primary* and *secondary* traditions, and say that secondary traditions contain and convey, in an abbreviated and perhaps emotionally bolstered form, information which could in principle, though perhaps only with a loss of convenience, be communicated also in a purely discursive fashion. The information embedded in primary traditions, on the other hand, cannot be separated from the way in which it is handed down, or rather it can be thus separated only within a context different in kind from that in which these traditions were originally functioning. In other words, secondary traditions can be dissolved without essentially impairing that activity whose transmission they serve; primary traditiones cannot. The thesis to the effect that there are primary traditions I shall call the *strong traditionalist* thesis, and contrast it with the *weak traditionalist* thesis which denies the existence of primary traditions but recognizes the existence, and usefulness, of secondary ones. The position denying this usefulness might then properly be called *anti-traditionalist*. I take the hard-core view of practical knowledge to imply, and be implied by, the strong traditionalist thesis. In what follows I will, very briefly, call attention to some of the issues bearing on this thesis; before doing that, however, I would like to touch upon two other, closely related topics.

The first is rationality. Reason and tradition are usually conceived of as opposed,[26] and even traditionalist arguments are often phrased in such a way as to maintain this opposition. The power of the irrational – or of the arational – is stressed, along with the importance of traditions as creating a dimension of coherence in the non-rational realm, as bringing, through their very irrationality, cohesion into society. It is in this sense that Karl Popper, quite a traditionalist in his way, writes:

What we call social life can exist only if we can know, and can have confidence, that there are things and events which must be so and cannot be otherwise. – It is here that the part played by tradition in our lives becomes understandable. We should be anxious, terrified, and frustrated, and we could not live in the social world, did it not contain a considerabla amount of order, a great number of regularites to which we can adjust ourselves. The mere existence of these regularities is perhaps more important than their peculiar merits or demerits. They are needed as regularities, and therefore handed on as traditions, whether or not they are in other respects rational or necessary or good or beautiful or what you will. There is a need for tradition in social life.[27]

Now if the strong traditionalist thesis holds, this way of formulating the matter is misleading. For the thesis implies that reason itself is ultimately grounded in traditions, or, as Oakeshott eloquently puts it: "'Rationality' is the certificate we give to any conduct which can maintain a place in the flow of sympathy, the coherence of activity, which composes a way of living".[28] It will not do to regard rationality, as Feyerabend does, as "one tradition among many rather than a standard to which traditions must conform",[29] since this would still amount to an unjustifiable picking out of some single tradition. A formula is needed which will preserve our intuitive grasp of what "rational" amounts to, without however introducing any arbitrary criteria. I think Oakeshott comes close to finding such a formula when, after writing that "no conduct, no action or series of actions, can be 'rational' or 'irrational' out of relation to the idiom of activity to which they belong", he goes on to state that "an activity as a whole (science, cooking, historical investigation, politics or poetry) cannot be said either to be 'rational' or 'irrational' unless we conceive all idioms of activity to be embraced in a single universe of activity".[30]

But the author who, in my opinion, really pointed the way here, even if for sixty years no one seems to have embarked upon it, was Maurice Halbwachs, in his *Les cadres sociaux de la mémoire*.[31] "Reason", Halbwachs wrote, "is actually a striving to raise oneself from a narrower to a broader tradition, into which latter the memories not merely of one class, but those of all groups will fit. ... Reason faces tradition as a broader society faces a narrower one".[32] The tradition capable of absorbing a variety of other traditions, or the tradition that emerges as an amalgam of various particular ones, will then possess, or amount to, what might be called *relative rationality*; and of course all rationality is relative.

The second topic I feel should be touched upon in the present context is the relation between traditionalism and the philosophy of mind. It seems to me that the strong traditionalist thesis is simply incompatible with what is usually called mentalism or intellectualism: the view of an autonomous, sovereign mind, of a mind intimately acquainted with, and freely operating upon, its own contents (images, concepts, and the like), a mind for which language, in particular, is a mere instrument of communication, an external vehicle expressing, and indeed guided by, inner thought-processes.

Wittgenstein and Ryle are of course well-known critics of this view, but their arguments are seldom taken notice of by traditionalist writers, generally insensitive to the epistemological presuppositions and implications of their position. Two notable exceptions were Edmund Burke and T.S. Eliot, who did indeed realize those implications. In his essay "Tradition and the Individual Talent", Eliot wrote:

The point of view which I am struggling to attack is perhaps related to the metaphysical theory of the substantial unity of the soul: for my meaning is, that the poet has, not a "personality" to express, but a particular medium, which is only a medium and not a personality, in which impressions and experiences combine in peculiar and unexpected ways. ... The emotion of art is impersonal. And the poet cannot reach this impersonality ... unless he lives in what is not merely the present, but the present moment of the past...[33]

And as to Burke, he not only had a theory of traditions, but in fact the rudiments of a theory of meaning to match the former. Examining the "common notion", according to which words "affect the mind by raising in it ideas of those things for which custom has appointed them to stand", Burke does "not find that once in twenty times" any such idea or "picture" is formed, and indeed when it is, "there is most commonly a particular effort of the imagination for that purpose". Burke gives a charming example. "Suppose", he writes,

we were to read a passage to this effect: "The river Danube rises in a moist and mountainous soil in the heart of Germany, where, winding to and fro, it waters several principalities, until, turning into Austria, and laving the walls of Vienna, it passes into Hungary; there with a vast flood, augmented by the Save and the Drave, it quits Christendom, and rolling through the barbarous countries which border on Tartary, it enters by many mouths into the Black Sea". In this description many things are mentioned, as mountains, rivers, cities, the sea, &c. But let anybody examine himself, and see whether he has had impressed on his imagination any pictures of a river, mountain, watery soil, Germany, &c. Indeed it is impossible, in the rapidity and quick succession of words in conversation, to have ideas both of the sound of the word, and of the thing represented; ... nor is it necessary that we should.

In the ordinary course of conversation, Burke concludes, "we are sufficiently understood without raising any images of the things concerning which we speak".[34] This is, clearly, an approach to meaning which does not presuppose or suggest mentalist views; it is compatible with the idea of language as an essentially social institution; it is, in particular, compatible with the strong traditionalist thesis.

Returning now to a brief examination of this thesis itself, we have to take into account, first of all, that the term "tradition" is surrounded by a family of related terms. This family would include terms like "authority", "convention", "custom", "disposition", "habit", "institution", "mentality", "mode", "mores", "norm", "paradigm", "practice", "prejudice", "rule", "style", "taste", "technique", The interconnections within this family are far from unequivocal, the meanings of most of the terms vary and overlap. Clearly, both a survey of connotations and a list of stipulations is called for.

Let us consider, for instance, the term "authority". According to Halbwachs, it is traditions which confer authority upon certain roles and persons.[35] Polanyi, on the other hand, stresses that only by *a previous act of affiliation*", by a "combined action of authority and trust". will the assimilation of basic traditions become possible at all.[36] Wittgenstein points out that one accepts very many beliefs on "human authority"[37] and seems to suggest a certain parallel between authority and tradition when declaring: "Tradition is not something a man can learn; not a thread he can pick up when he feels like it; any more than a man can choose his own ancestors".[38]

Or take the term "convention". For Hume and for Burke this notion was related rather than opposed to that of tradition. As Wilkins has put it:

Social conventions such as rules for the acquisition and transmission of property are artificial in the sense of being man-made, but given man's social nature and the mutual dependence of men there is a sense in which they are natural as well. The important thing for understanding both Hume and Burke is their general refusal to equate artificial with arbitrary.[39]

In a rather different context, in the domain of the philosophy of science, Fleck, too, strives to show that the element of arbitrariness has no primary role to play in the connotation of the term "convention", stressing "how little such conventions, which from the point of view of logic may seem equally possible, are in fact felt to be of equal value".[40] And in the domain of the philosophy of art it is e.g. Arnold Hauser who draws a close terminological parallel between convention and tradition. "Spontaneity and convention, originality and tradition", he writes, are

inseparable from each other... Every work, every form, and even the minutest attempt at expression ... are always the result of a conflict between spontaneity and convention, originality and tradition... The process is not one in which spontaneous personal experiences become communicable and accessible only through conventional forms, but one in which the experiences to be depicted move from the outset along conventionally regulated lines. ... Artistic expression comes about not in spite of, but thanks to, the resistance which convention offers to it.[41]

Clearly, Hauser is a strict traditionalist as far as the issue of artistic creativity goes, but it is the term "convention", not the term "tradition", that carries the weight of his argument. The connotations of "convention" are however no less blurred than those of "tradition". And here, most modern authors would seem to agree with Halbwachs e.g., for whom convention means *free agreement*: he contrasts the "purely conventional" with the "purely traditional".[42]

Or consider, again, the next term on our list, "custom". It is a term extremely rich in meanings. Burton Leiser in his book on the subject lists at least nine main ones, ranging from mere *habits*, through sanctioned *regulations*, to so-called *constitutive rules*, rules which, by their very definition, could not be broken.[43]

It appears that the terms surroundig "tradition" – the terms through the meanings of which the meaning of the latter could be explicated – themselves stand in need of elucidation. Obviously, then, a nominal explanation of the concept of tradition, though necessary, is not sufficient. What we need is not so much definitions, as much rather a detailed examination of the ways in which traditions, in all their forms and varieties, function at the different levels and in the different spheres of social life. Such spheres are: language, science, art, law, politics, education, and beyond them, or common to them, general phenomena like spontaneous orders, deviance and normality, creativity, group behaviour, and so on. Also the issue of so-called national or ethnic traditions, as well as the culture/civilization contrast would, in particular, merit special attention. – Here there already exists a substantial body of important research upon which one can draw. And I think much of that research directly supports the strong traditionalist thesis as formulated above. Thus with all the recent stress on linguistic universals and on the biological foundations of language, there has not survived, in the literature, any serious attempt to question the existence of essential linguistic layers culturally structured and traditionally transmitted. Noam Chomsky's oddly impoverished notion of linguistic creativity,[44] a creativity determined by genetic inheritance and following inborn patterns, has become a curio of the past. In a 1982 study Slobin and Bever could, once more, revert to Bloomfield's classic dictum "We speak ... by certain well-practiced schemes, – sentence-skeletons that require but the variation of a few words from utterance to utterance", and point to the language-specific nature and broad contextual setting of "schema-development".[45]

With respect to science, the role of traditions is an issue which, due to the Popper–Oakeshott controversy,[46] and especially to the controversy surrounding Kuhn's work,[47] has recently received ample attention. Important here is David Hollinger's observation that Kuhn has in fact applied to the history of science *the* conventional historiographic view of the part played by traditions in politics, arts, and the life of society in general.[48] "Kuhn's notion of the 'paradigm', his most celebrated and maligned term", writes Hollinger,

embodies the sense that activities are defined and controlled by tradition, and that tradition consists of a set of devices, or principles, that have proven their ability to order the experience of a given social constituency. An operative tradition provides a community with criteria to distinguish one activity from another, sets priorities among those activities, and enables the community to perform whatever common activities make it a community at all. Insofar as the community's common experience is contingent, that experience presents itself as a series of "problems" to be solved by tradition, which validates itself by transforming the contingency of experience into something comprehensible and subject to maximum control. Tradition, then, is socially grounded, and its function is that of organization. Organization may be achieved through a number of modes and devices, ranging from formal institutions to informal habits and from codes of abstract principles to concrete examples of how problems of a given class have been solved in the past. Whether it is conduct or perceptions that require organization, whether the task is prescriptive or cognitive, the organizing devices have enough flexibility to sustain them through successive, contingent experiences: to the extent that a tradition can expand and adapt, like the English common law, it is that much more likely to retain its constituency.[49]

Thus, as Hollinger on the other hand points out, in different communities – of which the community of modern-age natural scientists is only one specific kind – the role played by traditions may vary widely. Kuhn himself has written an essay in which he draws attention to the particular way traditions function in art, as contrasted with science. In art, but not in science, Kuhn emphasizes, a tradition might be dead yet its products still living; or again, "though resistance to innovation is a characteristic common to both art and science, posthumous recognition recurs with regularity only in the arts".[50] Also, artists "can and sometimes do voluntarily undertake dramatic changes in style on one or more occasions during their lives", whereas such changes are rare, and never voluntary, in the career of the individual scientist.[51] Still, not only will "most artists begin by painting in the style of their masters",[52] but one can also assume, Kuhn suggests, that even if styles might not, pictures do indeed serve as genuine "paradigms" in art.[53]

Mention has been made above of the traditionalist theory of art of Arnold Hauser. Again and again Hauser emphasizes that "every artist expresses himself in the language of his predecessors, his models, and his teachers", that "every newly created work owes more to other works than to the invention and experience of its creator".[54] Wittgenstein expresses a similar view when he says that "every composer changed the rules, but the variation was very slight; not all the rules were changed. The music was still good by a great many of the old rules".[55] According to Robert Musil even the spontaneity of an artist is inconceivable without handed-down forms and concepts – it is those very handed-down forms that become a source of originality in the creative process.[56] Or, to quote Hauser again:

what is most significant is not the fact that every expression uses conventional forms from the very beginning but the fact that conventional forms of expression themselves create in part the content of what is being expressed. ... It is true that expression always moves on well-worn tracks, but the tracks multiply and bifurcate as they are being traveled.[57]

And a related position has been developed, perhaps surprisingly, by Karl Popper. According to a "conjecture" of his

it was the canonization of Church melodies, the dogmatic restrictions on them, which produced the *cantus firmus* against which the counterpoint could develop. It was the established cantus firmus which provided the framework, the order, the regularity that made possible inventive freedom without chaos.[58]

In theories of law, politics, and of social life in general – theories in which fetishist categories like truth and beauty never really played a role – the idea of an order imposed by mere traditions of course always had a relatively stronger appeal. The words of Carl Menger, the inaugurator of the trend that has become known as the Austrian School of Economics, might convey a suggestion of the unlikely parallels here obtaining between Anglo-Saxon and German thought. Menger definitely displays strong sympathies towards views such as those of Burke, and of F. K. v. Savigny (chief representative of the German historical school of legal theory). As Menger writes:

Burke was probably the first, who, trained for it by the spirit of English jurisprudence, emphasized with full awareness the significance of the organic structures of social life and the partly unintended origin of these.[59]

Burke taught that numerous institutions of his country

were not the result of positive legislation or of the conscious common will of society
directed toward establishing these, but the unintended result of historical development. He first taught that what existed and had stood the test, what had developed
historically, was again to be respected, in contrast to the the projects of immature
desire for innovation. Herewith he made the first breach in the one-sided rationalism
and pragmatism of the Anglo-French Age of Enlightenment.[60]

There is, Menger maintains, a "subconscious wisdom" manifested in
those institutions that come about organically; and the meddlesome
advocates of reform "would do well less to trust their own insight and
energy than to leave the reshaping of society to the 'historical process of
development'".[61] – In a similar spirit, to-day's leading exponent of the
Austrian School, F. A. von Hayek stresses that "since we owe the order of
our society to a tradition of rules which we only imperfectly understand,
all progress must be based on tradition".[62] But the grand old man of
contemporary German philosophy, Hans-Georg Gadamer, too, realizes
that the ordering of life through the rules of law and morality always
amounts to more than the application of general principles. "Our
knowledge of law and morality", he writes, "is always supplemented from
the individual case, even productively determined by it. The judge does
not only apply the law in concreto, but contributes through his very
judgment to the development of the law".[63] And in this domain of theory,
too, the ideas of the later Wittgenstein have provided new impetus. It was
partly under his influence that H. L. A. Hart has developed his conception
of law as a combination of "primary" and "secondary" social rules. Hart's
primary rules seem to be a proper subclass of the primary traditions we
described above. They are customs supported by strong social pressure,
coming into being through "the slow process of growth, whereby courses
of conduct once thought optional become first habitual or usual, and then
obligatory...".[64] Without their prior existence, no legal system could be
built up.

When led by a sense for primary traditions, the discriminating eye will
soon find faults with much of the prevailing views on education. Here
again, the writings of T.S. Kuhn have shed new light on some crucial
problems. In having developed the truly revolutionary notion of normal
science, Kuhn underscored the need for rigid traditions within particular
scientific groups. In the absence of such traditions, scientific innovation
appears neither structurally nor psychologically possible.[65] This view has

immediate consequences for educational theory. As Kuhn has pointed out, scientific progress is, at least in the basic sciences, not achieved by "liberal" education, by encouraging "divergent" thinking.[66] And one can add that, at the elementary level, all learning seems to require a measure of external rigidity. It was Wittgenstein who, in his later philosophy, has explored the reasons for this, and it is significant that an elementary spelling book, his *Wörterbuch für Volksschulen* of 1926, was the prelude to that philosophy.[67] In spelling, as in elementary mathematics, Wittgenstein believed in authoritarian teaching methods. That these beliefs, rooted in sentiment but based on analysis, were far from being erroneous, to-day clearly emerges from a number of educational surveys and reports.[68]

Theoretical attitudes on deviance are in many ways bound up with those on education, and it is to be expected that an awareness of the essential organizational role more or less rigid traditions play in human communities will, again, preclude the acceptance of radically permissive sociological arguments.[69] But that very awareness – the conviction that only a social fabric entirely destroyed can be devoid of traditional elements – will also enable one to see through the claims of excessive traditionalism, to recognize invented traditions for what they are, to withstand the romantic yearning for bonds. Nationalism on the one hand, and the attacks on contemporary "civilization" in the name of some more authentic "culture" on the other, are two notable instances of an excessive traditionalist ideology. National divisions and nationalist sentiments are invariably bound up with material conditions surrounding the living. "Instead of being automatically united by a shared history", K. W. Deutsch wrote, "men ... cannot share the historical events through which they live, unless they are already in some sense united".[70] Nationalism as often as not forfeits the politico-economic present while focussing on an imagined past. Similarly, the foe of "civilization", while yearning for the fictitious warmth of an age that never existed, is blind to the real traditions of the society, the actual form of life, surrounding him. A seldom-quoted remark by Wittgenstein seems to be appropriate here. "It is very remarkable", he wrote in 1946, "that we should be inclined to think of civilization – houses, trees, cars, etc. – as separating man from his origins, from what is lofty and eternal, etc. Our civilized environment, along with its trees and plants, strikes us then as though it were cheaply wrapped in cellophane and isolated from everything great, from God, as it were. That is a remarkable picture that intrudes on us".[71]

CHAPTER 6

"TRADITION" AND RELATED TERMS:
A SEMANTIC SURVEY

The word "tradition" derives from the Latin term *traditio* ("delivery", "surrender", "handing over").[1] The various uses of *tradition*, and of the verbal form *tradere*, can be conveniently studied in the Bible. In the Vulgate there are some 500 occurrences of the term.[2] Some notable examples are: "Omnia mihi tradita sunt a Patre meo" ("All things are delivered unto me of my father", Mt 11, 27), "Proprio Filio suo non pepercit, sed pro nobis omnibus tradidit illum" ("He that spared not his own Son, but delivered him up for us all", Rom 8, 32), "Tradidi enim vobis in primis, quod et accepi" ("I delivered unto you first of all that which I also received", 1 Cor 15, 3), "Relinquentes enim mandatum Dei, tenetis traditionem hominum" ("laying aside the commandment of God, ye hold the tradition of men", Mk 7, 8), "Proficiebam in Iudaismo supra multos coaetaneos meos in genere meo, abundantius aemulator existens paternarum mearum traditionum" ("profited in the Jews' religion above many my equals in mine own nation, being more exceedingly zealous of the traditions of my fathers", Gal 1, 14). In the terminology of the Early Fathers *traditio* comes to mean the authoritative infallible preaching of the Church[3] – a usage in accordance with 2 Thess 2, 15: "Therefore, brethren, stand fast, and hold the traditions which ye have have been taught, whether by word, or our epistle". There occurs a shift during the Reformation. For Wiclif, Luther and Calvin only the Bible had authority, *traditiones humanae* were considered invalid, useless. As a reaction, the Catholic side stressed the importance of handed-down teachings not present in the Bible. The resulting, widespread, tendency was to exclude written, as opposed to oral, transmission from the meaning of the term. Samuel Johnson's definition of the English word *tradition* fully reflects this tendency.

Johnson lists[4] the meanings "The act or practice of delivering accounts from mouth to mouth without written memorials; communication from age to age" (illustrated by a quote from Hooker: "To learn it, we have tradition; namely, that so we believe, because both we from our predecessors, and they from theirs, have so received"), and "Any thing delivered orally from age to age" – the latter sense illustrated by the telling lines of

Milton, "They the truth / With superstitions and traditions taint". According to the 1864 revised edition Webster dictionary *tradition* means "The act of delivering into the hands of another; delivery", also "The unwritten or oral delivery of opinions, doctrines, practices, rites, and customs, from father to son, or from ancestors to posterity; the transmission of any opinions or practice from forefathers to descendants by oral communication, without written memorials", and "Hence, that which is transmitted orally from father to son, or from ancestors to posterity; knowledge or belief transmitted without the aid of written memorials". As the theological meaning of the term the Webster gives "That body of doctrine and discipline, or any article thereof, supposed to have been put forth by Christ or his apostles, and not committed to writing". Some main meanings given by the *OED* are "The action of handing over (something material) to another; delivery, transfer". "Delivery, *esp.* oral delivery, of information or instruction. Now *rare*", and "The action of transmitting or 'handing down', or fact of being handed down, from one to another, or from generation to generation; transmission of statements, beliefs, rules, customs, or the like, esp. by word of mouth or by practice without writing. Chiefly in phrase *by tradition*". And as a "more vague" sense the *OED* renders: "A long established and generally accepted custom or method of procedure, having almost the force of a law; an immemorial usage; the body (or any one) of the experiences and usages of any branch or school of art or literature, handed down by predecessors and generally followed".[5]

Modern usage, it appears, does not systematically distinguish between traditions on the one hand, and customs, conventions and the like on the other. Thus for instance that great foe of traditionality, Nietzsche, relatively seldom uses the expression "das Herkömmliche" (his favoured term for tradition).[6] He speaks of "customs", "conventionality", "convention", "fashion", and of *prejudice* ("opinions" of the "appropriated" or "public" kind, as he puts it).[7] H. B. Acton defines tradition as "a belief or practice transmitted from one generation to another and accepted as authoritative, or deferred to, without argument", and goes on to write: "it is clear that tradition and custom are closely connected, if not identical, notions, though we tend, perhaps, to use the word 'tradition' for the more elaborate and civilised forms of custom. A fuller treatment of them both would lead to the examination of such conceptions as those of habit and skill".[8] By contrast, Max Radin emphasizes that "only some of the inherited or transmitted customs, institutions, speech, dress, laws, songs and tales are traditions; and the use of the term implies a judgment about the value of the transmitted element".[9] D. M. Armstrong, too, finds that traditions are to be distinguished from "customs, habits, rituals, practices and rules for action", and stresses that the notion of tradition involves, as that of custom does not, a *normative* aspect.[10]

A precise definition of the term "tradition" will then, obviously, require more than just the mere recording of existing usage; it will involve explication and stipulation as well. And it seems that the first step must be a definition not of "tradition" itself, but of the less complex, though semantically interdependent, terms which constitute its background. A number of such terms have been already referred to in the foregoing. A more or less complete list would comprise ART (in the sense of skill), AUTHORITY, CONVENTION, CUSTOM, DISPOSITION, FASHION, HABIT, INSTITUTION, LAW, MANNER, MAXIM, MENTALITY, MODE, MORES, NORM, PARADIGM, PATTERN, PRACTICE, PREJUDICE, PRINCIPLE, RITUAL (RITE), ROUTINE, RULE, SCHEME, SKILL, STYLE, TASTE, TECHNIQUE, USAGE (USE), VALUE, and finally WAY (in the sense of manner, mode). These terms indicate specific contexts[11] in which certain regularities of behaviour can be interpreted. For a preliminary semantic analysis, let us here single out "art", "convention", "custom", "fashion", "habit", "institution", "prejudice", and "rule",

– "art": the nowadays most usual sense of this term, encompassing painting, architecture, etc., is of relatively recent date; the Latin and Greek words from which it is derived originally meant skill in joining or fitting. As a first meaning Johnson renders "The power of doing something not taught by nature and instinct; as, to *walk* is natural, to *dance* is an art", illustrated by a quote from South: "Art is properly an habitual knowledge of certain rules and maxims, by which a man is governed and directed in his actions". The second meaning he gives is "A science; as, the liberal *arts*", the third "A trade", illustrated by a quote from Boyle: "This observation is afforded us by the art of making sugar". Further meanings listed by Johnson are "Artfulness; skill; dexterity", "Cunning", and "Speculation". The Webster very clearly brings out the ambiguous relation between "art" and "science", giving, *inter alia*, "A system of rules serving to facilitate the performance of certain actions; – opposed to *science*, or to speculative principles; as, the *art* of building or engraving". then listing, however, the synonyms "Science; literature; aptitude;

readiness; skill; dexterity; adroitness", etc. Art in the sense of skill is, then, a habitual practice, learnt in accordance with customary ways of the given trade.

– "convention": derived from the Latin word *conventio* – meeting, assembly – the term primarily means *the action of convening*. Thus Johnson lists "The act of coming together; union; coalition; junction", "An assembly", and "A contract; an agreement for a time, previous to a definitive treaty". For "Conventional" he has "Stipulated; agreed on by compact". Of later date seem to be the connotations *standard behaviour* or *established way*. The entry in the Webster lists, *inter alia*, "General agreement or concurrence;[12] arbitrary custom; usage; conventionality; conventionalism", illustrating these meanings, significantly, by the lines of Tennyson: "There are thousands now / Such women, but convention beats them down". The two distinct senses spontaneity/compulsion are very clearly brought out by the Webster entry "Conventional": "Formed by agreement or compact; stipulated", and "Growing out of, or depending on, custom or tacit agreement; sanctioned by usage". Although in a rather more diversified context, basically the same contrast is conveyed by the *Oxford English Dictionary* where, for "convention", we have "General agreement or consent, deliberate or implicit, as constituting the origin and foundation of any custom, institution, opinion, etc., or as embodied in any accepted usage, standard of behaviour, method of artistic treatment, or the like. ... In a bad sense: Accepted usage become artificial and formal, and felt to be repressive of the natural in conduct or art; conventionalism", and "A rule or practice based upon general consent, or accepted and upheld by society at large; an arbitrary rule or practice recognised as valid in any particular art or study; a conventionalism".[13]

– "custom": the Webster here has "Frequent repetition of the same act; way of acting; ordinary manner; habitual practice; usage" – with an interesting quote from Raleigh: "Custom differeth from use as the cause from the effect, in that custom is by use and continuance established into a law" – and "(*Law.*) Long-established practice, or usage, considered as unwritten law, and resting for authority on long consent". And it lists, as *synonyms* for "custom": "Habit; usage; practice; fashion". Similarly the *OED*, which as the two first senses gives "A habitual or usual practice; common way of acting; usage, fashion, habit (either of an individual or of a community)" – quoting J. St. Mill: "The despotism of custom is everywhere the standing hindrance to human advancement" – and "*Law*. An established usage which by long continuance has acquired the force of a law or right, *esp.* the established usage of a particular locality, trade, society, or the like". In his book *Custom, Law, and Morality: Conflict and Continuity in Social Behavior* Burton M. Leiser stresses that the word "custom" covers a "family of meanings"; he mentions, *inter alia*, habits and routine, maxims or principles, style, rules of etiquette and of linguistic usage, rites and rituals.[14] And, as we have already seen, the term "custom" is especially closely related to the term "tradition". Thus for instance the American anthropologist Ruth Benedict can plausibly use the expression "traditional custom", e.g. when she speaks of "the formation of the individual's habit-patterns under the influence of traditional custom",[15] or indeed when she writes:

Traditional custom, taken the world over, is a mass of detailed behavior more astonishing than what any one person can ever evolve in individual actions no matter how aberrant. ... No man ever looks at the world with pristine eyes. He sees it edited by a definite set of customs and institutions and ways of thinking. ... The life-history of the individual is first and foremost an accomodation to the patterns and standards traditionally handed down in his community. From the moment of his birth the customs into which he is born shape his experience and behavior.[16]

– "fashion": deriving, ultimately, from the Latin word *facere*, "make", "do", this substantive, according to the *OED*, has or had the now rare or obsolete meanings "The action or process of making". "Make, build, shape; hence, appearance". "Form as opp. to matter". "Kind, sort". But it also meant or means "Manner, mode, way". "A current usage; *obs.* in *pl.* often = 'manners and customs', ways". and of course "Conventional usage in dress, mode of life, etc., *esp.* as observed in the upper circles of society; conformity to this".

– "habit": the first two senses of this term Johnson lists, "State of any thing: as *habit* of body", and "Dress; accoutrement; garment" recall some of the obsolete meanings of "fashion"; a third sense is conveyed by a near-quote from Locke, "Habit is a power or ability in man of doing any thing, when it has been acquired by frequent doing the same thing"; and a fourth is "Custom; inveterate use". The Webster has "The usual condition of a person or thing regarded as that which is had or retained; ordinary state, either natural or acquired; especially, physical temperament; as, a full, lax, or costive *habit* of body", "Fixed or established custom; ordinary course of conduct; hence, prominently, the involuntary tendency to perform certain actions which is acquired by their frequent repetition; as, *habit* is second nature; also, prevailing dispositions, feelings, and actions which are right or wrong; moral character", and "Outward appearance;

attire; dress", etc. As synonyms, "practice", "mode", "manner", "way", and "custom" are listed, and then there follows a short discussion of the relation of "habit" to "custom":

Habit is an internal principle which leads us to do easily, naturally, and with growing certainty, what we do often; *custom* is external, being habitual use or the frequent repetition of the same act. The two operate reciprocally on each other. The *custom* of giving produces a *habit* of liberality; *habits* of devotion promote the *custom* of going to church. *Custom* also supposes an act of the will, selecting given modes of procedure; *habit* is a law of our being, a kind of "second nature" which grows up within us.

To this the Webster adds two illuminating quotes, "Upheld by old repute, / Consent, or custom" (Milton), and "How use doth breed a habit in man" (Shakespeare). Now taking the presence of an act of will[17] as the feature distinguishing custom from habit is not, perhaps, of particular help; the internal/external distinction however, together with the churchgoer example, is useful, pointing, in fact, to the characteristically personal element in habit as contrasted with the rather more social nature of custom. This is the element John Dewey emphasizes in his *Human Nature and Conduct* (1922) when he equates customs with "established collective habits", or writes that "custom is essentially a fact of associated living whose force is dominant in forming the habits of individuals".[18] According-ing to the *OED*, the sense development of the term,

as seen in Latin and the modern languages taken together, is thus: *orig.* Holding, having, 'havour'; hence the way in which one holds or has oneself, i.e. the mode or condition in which one is, exists, or exhibits oneself, *a)* externally; hence demeanour, outward appearance, fashion of body, mode of clothing oneself, dress, habitation; *b)* in mind, character, or life; hence, mental constitution, character, disposition, way of acting, comporting oneself, or dealing with things, habitual or customary way (of acting, etc.), personal custom, accustomedness. This development was largely completed in ancient Latin, and had received some extension in O[ld] F[rench], before the word became English...

As the "most usual current sense", the OED gives "A settled disposition or tendency to act in a certain way, esp. one acquired by frequent repetition of the same act until it becomes almost or quite involuntary; a settled practice, custom, usage; a customary way or manner or acting", goes on however by pointing out that there is "no etymological ground for the distinctive use of 'habit' for an *acquired* tendency; but in philosophi-cal language, such a sense occurs already in Cicero". Thus arises a distinction, in the usage of philosophers, between "habit" and "disposition". The *OED* quotes Sir William Hamilton's explanation: "Both ... are tendencies to action; but *disposition* properly denotes a natural tendency, *habit* an acquired tendency".

– "institution": This term on the one hand seems to have a very broad meaning,[19] encompassing many of the related terms from "convention" to "rule"; on the other hand it somehow suggests a congealed pattern, a materialized structure, a social-physical *entity*. In this sense does Dewey speak of "institutions as embodied habits"[20]; or Edward Shils of the "minimal institutional setting" necessary for "the transmission of tradition", a setting like the *family*, or the "relationship of an apprentice to a master", or "institutions of elementary schooling", or even "ecclesiastical institutions, economic institutions, political parties, the state, and the system of stratification".[21] This seems to be the sense, too, which Waldenfels has in mind when he refers to rules as "the framework of action at a preinstitutional level",[22] or which J. G. A. Pocock thinks of when he writes about "extrapolations of institutional continuities" giving rise to the conception of traditions, or, on the other hand, about traditions that are *neither* "institutional", *nor even* "social".[23] And there is an intriguing definition by Max Radin – "an institution is a form of social organization which requires a functional classification of persons"[24] – into which, however, the important example of *language*, often regarded as the very paradigm of, or the paramount, institution, does not seem to fit.

– "prejudice": derived from the Latin word *praejudicium*, this term has undergone a radical change of meaning since classical times,[25] and has acquired a predominantly negative connotation.[26] However, there are some notable instances of a more affirmative interpretation of the term. Thus in one of his most often-quoted passages Edmund Burke wrote:

Instead of casting away all our old prejudices, we cherish them to a very considerable degree; and, to take more shame to ourselves, we cherish them because they are prejudices... Many of our men of speculation, instead of exploding general prejudices, employ their sagacity to discover the latent wisdom which prevails in them. If they find what they seek, (and they seldom fail,) they think it more wise to continue with the prejudice, with the reason involved, than to cast away the coat of prejudice, and to leave nothing but the naked reason; because prejudice, with its reason, has a motive to give action to that reason, and an affection which will give it permanence. Prejudice is of ready application in the emergency; it previously engages the mind in a steady course of wisdom and virtue, and does not leave the man hesitating in the moment of decision, skeptical, puzzled, and unresolved. Prejudice renders a man's virtue his habit, and not a series of unconnected acts. Through just prejudice, his duty becomes a part of his nature.[27]

But note here, again, the concluding reference to "just" prejudice, with its implication that not all prejudices are just. And it is of course the idea of the unjust, the malign, prejudice which constitutes the generally accepted meaning of this term. It is in this sense that the arch-liberal Austrian physicist Ernst Mach could speak of "the fetters of inherited prejudice", or of the "terrible power" of what we call – as the translation puts it – "prejudgment or prejudice", i.e. "habitual judgment, applied to a new case without antecedent tests". But even Mach concedes that without certain "fixed habitudes of thought", new problems would not become perceivable at all. "No one could exist intellectually", Mach writes,

if he had to form judgments on every passing experience, instead of allowing himself to be controlled by the judgments he has already formed. ... On prejudices, that is, on habitual judgments not tested in every case to which they are applied, reposes a goodly portion of the thought and work of the natural scientist. On prejudices reposes most of the conduct of society. With the sudden disappearance of prejudice society would hopelessly dissolve.[28]

And it was in this spirit that Robert Musil wrote:

In his potentialities, plans, and emotions, man must first of all be hedged in by prejudices, traditions, difficulties and limitations of every kind, like a lunatic in his straitjacket, and only then will whatever he is capable of bringing forth perhaps have some value, solidity and permanence.[29]

– "rule": the term is derived from the Latin word *regula*, meaning a ruler, rule, model. The philosophically central meaning, as conveyed by one of the Webster's definitions, is: "That which is prescribed or laid down as a guide to conduct; that which is settled by authority or custom; a regulation; a prescription; a minor law; a uniform course of things".[30] The *Shorter Oxford English Dictionary* has, *inter alia*, "A principle, regulation, or maxim governing individual conduct". "A principle regulating practice or procedure". "Without article: Rigid system or routine". "A formal order or regulation governing the procedure or decisions of a court of law; an enunciation or doctrine forming part of the common law, or having the force of law". "A standard of discrimination or estimation; a criterion, test, canon". – To *follow a rule*, then, means more than merely to act in a regular fashion. As Durkheim has put it, a principle becomes a "rule of conduct" only if "the group consecrates it with its authority". Accordingly: "A rule, indeed, is not only an habitual means of acting; it is, above all, *an obligatory means of acting*; which is to say, withdrawn from individual discretion". [31] A detailed analysis of the relation between "rule" and "habit" is given in H. L. A. Hart's *The Concept of Law*. "What *are* rules? What does it mean to say", Hart asks, "that a rule *exists?*" The answer he ultimately arrives at is that

there is involved in the existence of any social rules a combination of regular conduct with a distinctive attitude to that conduct as a standard. ... [A] varied normative vocabulary ('ought', 'must', 'should') is used to draw attention to the standard and the deviation from it, and to formulate the demands, criticisms, or acknowledgements which may be based on it.

By contrast, for a group "to have a *habit* it is enough that their behaviour in fact converges. Deviation from the regular course need not be a matter for any form of criticism".[32]

Now these terms, and some of the related ones not yet explicitly discussed, could be construed as forming a kind of system. Drawing up this system, let us first consider the term "convention", the lexicographic description of which distinctly suggested two opposing meanings, namely *arbitrary* convention on the one hand, and *inherited/compulsory* convention on the other. Here the obvious solution is to use in fact two different expressions, say convention$_{ARB}$ and convention$_{INH}$. We can then, for instance, recognize a basic theoretical agreement between the views of the French sociologist Maurice Halbwachs who contrasts the "purely conventional" with the "purely traditional",[33] and those of the Hungarian philosopher of art Arnold Hauser, who draws a close terminological parallel between convention and tradition.[34] Clearly, Halbwachs here speaks about convention$_{ARB}$, and Hauser about convention$_{INH}$.

As a second step, we regard the set "custom", "fashion", "habit", "law", "practice" and "usage". Earlier our attention has been drawn to the fact that customs can plausibly said to be social habits. One can also maintain that whereas customs are socially *sanctioned*, habits are not. Now there are different degrees of sanctions: mere *criticism*, signalling deviance from the standard; diffused *social pressure*; and *organized sanctions* enforced by a legal system. Customs can be, or can become, laws; but not all customs are laws, and we might say, quite plausibly again, that while (customary or statutory) laws are enforced by sanctions, customs are enforced by social pressure. And we stipulate that fashion, practice, and usage constitute standards against which deviations can be criticized, but are such that conformity with them will not be enforced. The term "taste" we again take to have *two* different senses: following

Burke, we speak of *natural* and *acquired* taste,[35] introducing the terms taste$_{NAT}$ and taste$_{ACQ}$. Taste$_{ACQ}$ we classify with fashion.[36] We thus have the sequence HABIT – FASHION/PRACTICE/ TASTE$_{ACQ}$/USAGE – CUSTOM – LAW, arranged along what might be regarded as the dimension of sanctions. As a further stipulation we take "rule" and "norm" to mean *social* rule and *social* norm, i.e. to play roughly the same role as "custom", and finally add to the triad CUSTOM/ NORM/RULE, as a fourth element, CONVENTION$_{INH}$.

Thirdly we take habit to be acquired, disposition to be natural or inborn, and express this by saying that for a habit to have come into existence a certain interval of *social time* must be presupposed, while for dispositions that interval is zero. Since habits pertain to individuals the time required for their acquisition should not exceed the life span of a single generation; by contrast, we stipulate that *mentality* is a socially inherited habit, the existence of which, then, presupposes a multiple-generation interval in social time, i.e. an interval stretching, at least, across two generations. We classify taste$_{NAT}$ with disposition, skill on the one hand and prejudice on the other with habit,[37] and arrive at the sequence DISPOSITION/TASTE$_{NAT}$ – HABIT/PREJUDICE/SKILL – MENTALITY, ordered along the dimension of social time. On the other hand we can plausibly maintain that convention$_{ARB}$ needs no extended social time either, being, in this respect, similar to disposition; once agreed upon, however, it will serve as a standard, and is bound up, in this sense, with (a minimum measure of) sanctions. (Compare Diagram 1.)

The family of terms surrounding "tradition" all of them designate attitudes, tendencies, or frameworks of action that display, or are conducive to the establishment of, regularities or uniformities. They more or less imply, also, the absence of the unprecedented, of alternatives, of deliberation, of reflection. The reflective attitude is expressed by another family of terms. This family would include expressions like CALCULA- TION, DEDUCTION, DELIBERATION, INFERENCE, JUDGEMENT, RATIOCINATION, RATIONAL, REASONING, RECKONING, REFLECTING, and perhaps SPONTANEOUS, in the sense of "free", "of one's own accord".

While the former family consists exclusively of substantives, most of the members of the latter one can also be expressed as verbal substan- tives, or indeed verbs. "Rational" and "spontaneous" are adjectives, but their adverbial forms are just as natural. The first cluster refers to frameworks of activities; the second to activities that are characteristically supposed not to require a framework. The argument of the present chapter is influenced, in contrast, by the supposition that reflection invariably proceeds along socially acquired lines. As Dewey has put it, "an idea gets shape and consistency only when it has a habit back of it" – thinking, if it is "not a part of ordinary habits ... is a separate habit alongside other habits". The "reflective disposition arises in some exceptional circumstance out of social customs", "the rational attitude" is an acquired "disposition", reason is "a laborious achievement of habit".[38]

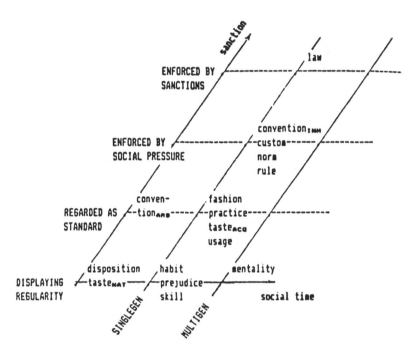

Diagram 1. The family of terms surrounding "tradition" is placed into a two-dimensional space, defined by the axes *sanction* and *social time*.

Between the non-reflective attitude on the one hand and the reflective one on the other, traditions seem to occupy an intermediary position.[39] Traditions appear as non-reflective in the sense that we would find the idea of someone deliberately *choosing* a tradition, or of a basically critical attitude towards a tradition one *has*, to be almost a contradiction in terms. This fundamental aspect of tradition comes to the fore in that stress on oral, as contrasted with written, delivery one encounters in the earlier connotations of the term. The process of initiation into an oral tradition means inculcation, involves drill, a kind of passive reception on the part of the novice, that will leave no scope for the distance, interpretation or criticism a tradition handed down in writing would render possible. Hence the emphasis on reading the Scriptures in Protestantism.[40] Hence the

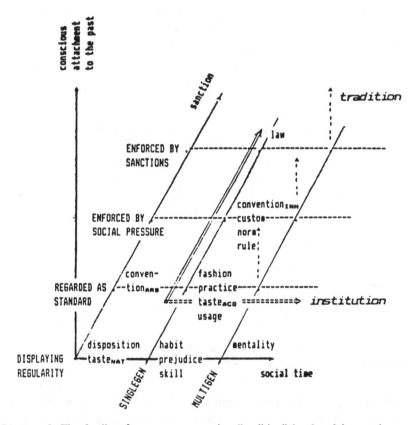

Diagram 2. The family of terms encompassing "tradition" is placed into a three-dimensional space, defined by the axes *sanction, social time,* and *conscious attachment to the past.*

broad interpretative freedom the US Supreme Court has in comparison to the British House of Lords: the former, relying on a Constitution as the latter does not, "can always abandon what has been said in order to go back to the written document itself".[41]

On the other hand traditions do rely upon reflection in the sense that they fundamentally involve a kind of *consciousness of the past*. In the formulation of Max Radin, *recollection* "seems to be the essence of tradition. ... [an] institution becomes traditional when it is recalled that it existed in past times and when at least some persons now desire to continue it".[42] Halbwachs indeed identifies traditions with collective memories.[43] Josef Pieper spoke of tradition and of recollection as interrelated notions: "Erinnerung, hier als ein kollektiver Prozeß gedacht, und Überlieferung haben das gemeinsam, daß in beiden der Ursprung anwesend gehalten und vergegenwärtigt wird".[44] Edward Shils wrote, in his classic paper of 1958, of an "affirmative attachment to the past",[45] Pocock of the "knowledge", or at least "assumption", of a tradition's indefinitely ancient history.[46] Armstrong, too, brings out the element of consciousness pertaining to tradition when he writes that the latter seems to be "a peculiarly human thing: a mark of the rational animal", since its enactors "almost inevitably must have a language, one that can be used to recommend the course of conduct involved in the tradition".[47]

We may arrive, then, at a plausible explication of the term "tradition" by restricting the modern, for the purposes of theoretical analysis much too broad, usage: by retrieving from the older sense the element "handed down *from generation to generation*",[48] and by stressing the aspect of active recollection, of conscious attachment, of "receptive affirmation" (Shils).[49] We will designate, then, by the term "tradition" any such practice, custom, etc., which is accepted as authoritative, requires conscious adherence, the history of which extends over at least three generations,[50] and which is known by its adherents to have that history (see DIAGRAM 2). *Fictitious traditions*, traditions having a largely or purely imaginary content or assuming unreal time dimensions, do not necessarily fall, according to this explication, outside the boundaries of traditions proper.[51] – As regards receptive affirmation, or conscious attachment to the past, the transition between practices/customs/laws on the one hand, and traditions on the other, is taken to be a continuous one; in other words, adherence to customs etc. might involve, though it does not need to involve, a measure of historical consciousness. The notion of such consciousness should not, however, form part of the meaning of "institution". Institutions we stipulate to be the strictly here and now side of practices, customs, laws, etc. – frameworks for ongoing activities devoid of memory.

CHAPTER 7

HISTORICAL CONSCIOUSNESS IN THE COMPUTER AGE

That there exists no healthy society that is without some sort of historical memory is an assumption predominating both in the popular mind and also among social theorists. Another formulation of this assumption might read: it is part of the normal present of a society to have a consciousness of its past; in the absence of common memories a society as it were falls apart, for there no longer obtains the necessary solidarity among its members.[1]

Standing opposed to this assumption is what could be called the post-modern view – a view I will here defend with some crucial qualifications – according to which society is held together not by any special *contents* of consciousness, and therefore also not by a consciousness of some common past, but rather through the traffic of its members one with another, through the exchange of goods, services, and especially of *information*. Society is a matter of *communication*,[2] and on the view in question the framework of communication need not at all contain a dimension referring to the past. Indeed it is increasingly affirmed precisely of our own age, the age of information, of electronic, audio-visual, computer-driven communication, that it does not need such a dimension. My aim in what follows, then, is to subject this assertion to a sympathetic, but somewhat closer, inspection. First, however, some characteristics of the *earlier* phases of the technology of human communication have to be considered.

The history of preserving and of passing on the knowledge of society – the history of the technology of communication – can be divided into four main phases:[3]

1. the phase of primary orality,
2. the phase of literacy,
3. the typographic phase (printing),
4. the phase of what has been called "secondary orality", of electronic information processing and transfer.

A culture is said to be one of primary orality if it does not yet possess the technology of writing at all, or only in a very rudimentary fashion – as *aides-mémoire*, for example in the case of the Peruvian *quipus* or knot language. Words, in such a culture, are exclusively spoken or heard; the knowledge society possesses has to be stored in easily memorizable formulae; it has to be memorized through constant repetition of authoritative texts – it has to be passed on through *tradition*.

It belongs to the nature of traditions, now, that their truth-content is not allowed to be called into question. Doubt would, after all, destroy precisely that immediate and devoted acceptance of the hearer without which the committing to memory of what is heard can come about only imperfectly. The indubitability of the transmitted text is legitimated through the fiction that this same text has been passed on *unchanged from generation to generation* – leading back to what is in the end a divine origin. And this same resistance to doubt is reinforced through elements of solemnity and ritual.[4]

In fact, however, an unchanged passing on of tradition is quite dysfunctional, is indeed such as to influence negatively the capacity of society to adjust itself to new circumstances. If the flexibility of tradition – which is to say of the oral preservation and transmission of knowledge – is hampered, e.g. through its becoming partially fixed in a rudimentary pre-alphabetical sign-system, or perhaps through ways which give rise to the dominance of a rigid educational elite, then there arises a cognitive blindness to change, a blindness which leads finally to the collapse of the culture in question.[5]

A *functional* tradition, in contrast, is a homeostatic affair: its content is adjusted, constantly and automatically, to what is taking place in the present.[6] This means also, however, that traditions can convey no image of the past which is objective in the modern sense of this word, indeed that they know no historical past at all. Eric Havelock writes of "a present tradition extending into the past and expected to extend into the future: The idiom in which the three periods are described establishes their identity, not difference".[7] The individual in a non-literate society, write Goody and Watt in their classic study, "has little perception of the past except in terms of the present".[8] Traditional knowledge is ahistorical, *blind to what is past.*

The technology of writing, on the other hand, which for the first time allows the fixing of what is said and the comparison of different texts, leads to the question of the strict identity or difference of utterances and thus to the idea of contradiction and of coherence, to critical-rational thought. It is the appearance of literacy, and especially the emergence of Greek alphabetic writing, which for the first time allows a precise *objectivized* representation of spoken thought. It thereby makes possible that *distance* of the cognitive subject to its own mental contents, that intellectual space, in which conceptuality and reflection can for the first time unfold themselves. Only with the appearance of writing does there occur the differentiation of legend and fact, myth and knowledge, and the beginning of a feeling for history and for historical *distance*. The transition from orality to literacy is from the point of view of the theory of knowledge quite decisive, as has been recognized in a rapidly growing literature since the 1960s.[9]

Already in the work of Oswald Spengler, however, one finds the idea that writing is a quite new type of language, implying "a complete change in the relations of man's waking consciousness", in that writing liberates consciousness "*from the tyranny of the present*". Speaking and hearing, Spengler writes, take place only in proximity and in the present, writing "*is the grand symbol of the Far*", and not only of distance in space, but also of duration.[10]

The full unfolding of this new idea occurred however only with the appearance of the printed book. The age of manuscript culture remains still an overwhelmingly oral period; texts are hard to produce and expensive to obtain; they are able to serve as an aid to oral communication, not, however, as its substitute. Not only in the Middle Ages but also in the early modern period, then, the "space of historical experience" of Europe is marked by "a deep feeling of unity spanning generations", as Reinhart Koselleck writes.[11] And this is connected not only with the absence of "revolutionary" technical developments, and with the fact that, generally speaking, historical events were experienced as familiar, recurring – factors mentioned by Koselleck – but also, and above all, with the technology of communication then still dominant: the orality of the manuscript culture and of the early book culture[12] – a technology which *allows no separation of the generations within the process of learning*.[13]

The age of print, in contrast, creates not only the possibility of independent learning, but also, with its wealth of books, creates reliable and constant texts. With its standardized chronologies and taxonomies it creates the possibility of a unitary science, of cumulative and critical knowledge, of the idea of progress, and of our modern historical consciousness. As Elizabeth Eisenstein in her *The Printing Press as an Agent of Change* observes:

More abundantly stocked bookshelves obviously increased opportunities to consult and compare different texts. Merely by making more scrambled data available, by increasing the output of Aristotelian, Alexandrian and Arabic texts, printers encouraged efforts to unscramble these data. Some medieval coastal maps had long been more accurate than many ancient ones, but few eyes had seen either. Much as maps from different regions and epochs were brought into contact in the course of preparing editions of atlases, so too were technical texts brought together in certain physicians' and astronomers' libraries. Contradictions became more visible; divergent traditions more difficult to reconcile. ... Before trying to account for an "idea" of progress we might look more closely at the duplicating process that made possible not only a sequence of improved editions but also a continuous accumulation of fixed records. For it seems to have been permanence that introduced progressive change.[14]

It was the preservation of the old, then, which launched a tradition of the new. A shift in communications preceded the rise of a modern historical consciousness by a century or more. "The past could not be set at a fixed distance", writes Eisentein, "until a uniform spatial and temporal framework had been constructed".[15]

Historical consciousness, the explicit consciousness of the fact that what is past is *essentially* different from what is present,[16] implies a fundamental scepticism as regards the usefulness of historical experience. The unfolding of historical consciousness and of the idea of progress is characterized by Koselleck in terms of a divergence of the "limits of the space of experience and the horizon of expectation". It becomes almost a rule, he writes, "that all previous experience can serve as no objection to the otherness of the future".[17] Koselleck cites a contemporary of the French Revolution, for whom that event was an experience which seemed to the world "to pour scorn on all historical wisdom... There grew out of it daily new phenomena to which one could find ever fewer parallels in history".[18] "Inherited historical experience", as Koselleck writes,

was no longer able to be extended directly to experience. ... To this there has come to be added since the end of the 18th century one further fact: that of technical-industrial progress... It becomes a general empirical theorem of scientific invention and of its industrial application that it allows progress to be expected, without however allowing it to be calculated in advance.[19]

Koselleck's thesis is

that in the modern age the difference between experience and expectation expands increasingly, or more precisely that the modern age can be conceived of as a new period in human history only since expectations have distanced themselves ever more from the experiences hitherto made... In virtue of this acceleration both political-social as also scientific-technical progress change the rhythm of time and the time intervals of the life-world.[20]

This ability to realize that the present does not resemble the past, the clear-sightedness acquired by modern man, is however bound up with an illusion of the Enlightenment and generates a certain anthropological blindness. For between the past and the present there do of course obtain deep continuities. Even in the modern life-world there exist pre-literal, non-literal, and even entirely extra-linguistic dimensions. The socialization of young children is pre-literal. The communication between the individual and his primary groups is non-literal (it is spoken, not written).[21] Entirely outside the sphere of language or very nearly so is the sphere of those technical objectivations, processes and skills of which István Hajnal could affirm that it is precisely in this sphere that "true inventions" take place.[22] Because writing creates a certain prejudice in favour of pure reason and thereby a certain distance to the oral and practical levels, it has the effect of alienating its users from the real life-world — an effect of which Plato was conscious, but to which, all the same, his epistemology fell victim. Literacy fully unfolded brings with it also a further sort of epistemological alienation. As Goody and Watt write:

the mere size of the literate repertoire means that the proportion of the whole which any one individual knows must be infinitesimal in comparison with what obtains in oral culture. Literate society, merely by having no system of elimination, no "structural amnesia", prevents the individual from participating fully in the total cultural tradition to anything like the extent possible in a non-literate society.[23]

It is from this perspective, of a lack of "social amnesia", that Goody and Watt make sense of Nietzsche's tendency

to describe "we moderns" as "wandering encyclopaedias". unable to live and act in the present and obsessed by a "'historical sense' that injures and finally destroys the living thing, be it a man or a people or a system of culture".[24]

It is clear that it is only the age of the printed book which is fully affected by this diagnosis. In Klaus Haefner's formulation, the current information explosion "with its daily increase of 10 million printed and published pages has forced modern man into a corner of his informational environment which seems relatively arbitrary".[25]

Three sorts of cognitive blindness manifest themselves here. First, there is the sort of blindness which follows from the epistemological effects of writing as such: the blindness to *practical* knowledge, to skill and dexterity. Second, there is the informational deficit of the individual in relation to the total knowledge of his community, or of the society

surrounding him, a deficit which arises with literacy and intensifies
progressively during the era of typography. Finally there is the incapacity
to discover what is common to the past and present and thus to draw
lessons from the past for the present. Historical consciousness, the typical
consciousness of the typographical culture, is in this sense *blind to the
present* – a blindness which, however, the *science* of history does not
necessarily have to share. Historical science may be something which is
marked by both a sense of history *and*, against this background, a
capacity to perceive the social structures and relations currently obtaining.[26]

What sort of changes are brought about here by developments in
computer technology? As regards the blindness concerning practical
knowledge, it was precisely the concern with computers in the context of
artificial intelligence research which led, in this connection, to a
breakthrough in philosophical understanding. That it is precisely our
common everyday abilities which are the most difficult to convey to a
computer is a discovery which today belongs to the generally accepted
wisdom of artificial intelligence research. Now as regards the informational chasm between the individual and his society, this is of course the
very problem the solution of which was the original task of electronic
data processing and networking. What, now, of the blindness to the
present that characterizes historical consciousness? What indeed is to be
said about historical consciousness as such in the newly unfolding age of
information?

The computer presupposes our literate culture. One types material in,
one reads what is on the screen; one processes texts and produces
printouts. It is asserted also that electronics does not inhibit but rather
expands the production of books.[27] And yet it is presumably correct to
conceive the age of information as a new world of orality. For it is first of
all the case that the fundamental technologies of the computer and of the
transmission and fixation of sound and pictures do overlap to a very great
degree. Secondly, the computer is involved directly in ordering the mass
of information that is broadcast on radio and television. And thirdly one
does after all hear that computers will gradually, albeit perhaps in a
rudimentary fashion only, become able to grasp our spoken language.

The world of primary orality was, now, oriented about the present and
blind to the past; from this however one should not immediately draw
conclusions in relation to what we have called above "secondary orality".[28] Contemporary orality, as Walter J. Ong emphasizes,

is post-typographical, incorporating an individualized self-consciousness developed with the aid of writing and print and possessed of more reflectiveness, historical sense, and organized purposefulness than was possible in preliterate oral cultures. ... One of the characteristics of our present culture, with its massive control of knowledge through electronic as well as typographic, chirographic, and oral media, is that it has situated man within his own history and thus given him a sense of self-possession previously unrealizable. ... Explicit and highly analytic knowledge of the past rather than the emotional commitment to ancestors, the "dead", gives man today his sense of identity.[29]

Yet one can still ask whether this picture drawn by Ong can claim a more than transient validity; or whether there are not certain tendencies at work in the information age which subvert and drive out the literal and typographical culture, and more particularly alter the corresponding consciousness of the past. One of the most immediate sociological consequences of the computer is indeed its quite considerable influence upon our sense of time. For if historical consciousness presupposes an environment in which historical changes take place quickly enough to make themselves felt at all, computerization seems in contrast to have brought about already a technological acceleration which is such as to make even the most recent past irrelevant and even the most imminent future radically unforeseeable. We are held as it were imprisoned in the present. Hermann Lübbe has considered this problem in a number of studies[30] and I would like to mention once more the information theorist and classical scholar David Bolter who finds it somewhat ironic "that progress in the computers and in the rest of our current technology is so rapid that it tends to negate history".[31] The internal manner of functioning of a computer too leads to a new experience of time and progress. On the one hand Bolter recalls under the influence of Spengler the ancient idea of a declining or *self-repeating* history, and on the other hand he points to Christian eschatology, remarking that the computer time is *finite* and the mode of operation of the computer is *cyclic*. "The experience of programming", he writes, "shared by millions of educated people, is helping to change our culture's view of progress and perhaps its view of history itself".[32] These observations are not, it is true, entirely convincing. That the internal time of the computer is other than our usual one, that it has a different *duration*, is however clear. Just think of the interesting concept of "realtime".

The phenomenon of word processing acquires here a special significance. The spoken word is flexible, elastic, but vanishes in the moment of speaking. Written language – and even more: printed language

– are enduring but rigid. A text that is stored in the computer, in contrast, is preserved, but also changeable. Language that is entrusted to a word processor is, as Richard Dimler writes, "dynamic rather than static, malleable rather than fixed, soft rather than hard, plastic rather than rigid. As a consequence language never seems to reach a finished stage".[33] One can of course date one's files, and specific printouts can be preserved for ever; these are however reflexes of the typographical culture. The text called up from the memory of the computer is as such always simultaneous, is lacking in all history. If it has been stored in an older system, then it will be converted – not, for example, translated or reconstructed. Age-old documents preserved in the computer carry no mark of temporality.

That an environment of timeless texts cannot remain without influence on our sense of time seems to me unquestionable. And we have concentrated here exclusively on the mere text processor – not for example on projects in the realm of artificial intelligence, or on the phenomenon of texts which can change themselves between one reading and the next.[34]

I have mentioned here matters which have to do with the computer as it really is. Before closing I want to refer briefly to that *ideal* of the emerging age of information which, even though it expresses *an illusion*, is yet not for that reason without effects. And in my opinion this ideal is able to contribute quite essentially to a subversion of the modern historical consciousness. I refer to the ideal of *omniscience*, the illusion that in a culture graced with intelligent computers and global networks everything is *possible*,[35] every problem *solvable*, every question *answerable*. Crucially affected by this ideal is the concept of *simulation*. Simulation relates to future events; yet still it must of course occur on the basis of present knowledge. The assumption that one can simulate social-historical processes, even over long time periods and with high probability, seems to come close to the conviction that one understands society and controls history. It comes down to an attitude according to which one could – in Koselleck's terminology – *experience an expectation*. This, however, is an attitude which is completely incompatible with modern historical consciousness. It conveys an illusory image of the relation of the future to the present. It is *blind to the future*.[36]

In presenting the disappearance of historical consciousness as a *danger* that is bound up with the new information technology, what I am aiming at is not some new-fangled sort of conservative cultural criticism. What concerns me is, rather, a sober scientific knowledge of history – and thereby of the still effective traces of the past – the very possibility of which in the future I perceive as being under threat. And nor would I allow myself any illusions concerning the possibility of warding off this threat, were it not for the fact that one can discern a quite definite contrary tendency precisely in this new technology itself. I refer, here, to the gradually dawning purely practical realization that information technology cannot after all be divorced from more classical forms of knowledge, that simple computer-English, too, needs the background of the study of English language and letters. The assumption that for example spelling or hyphenization errors might be capable of being corrected automatically – this assumption is misleading, as every printer knows. As the software-specialist Edsger W. Dijkstra writes: "Besides a mathematical inclination, an exceptionally good mastery of one's native tongue is the most vital asset of a competent programmer".[37]

That a solid semantic apparatus belongs to the presuppositions of clear thinking should of course be evident from the very start. Somewhat less trivial is the realization that the knowledge of classical languages can contribute crucially to the mastery of one's native tongue. This thesis was defended by Ernst Mach, someone who was otherwise inclined to favour the "new directions of progress", the "enlightenment that has come from the great mathematical and physical researches of the last centuries", over against the attachment to ancient cultures. Mach saw it as an obvious consequence of schooling in the ancient languages that it promotes "lucidity and accuracy of expression".[38] Clearly, however, the intelligent employment of ancient Greek or of Latin is hardly imaginable without a corresponding knowledge of the culture and history of the ancient peoples. But this then means that the information explosion of today and tomorrow cannot be mastered, even if we take electronic data processing for granted, by people who are not educated in the classical sense. The perfect mastery of information technology seems to be impossible without an "understanding of our cultural heritage".[39] This task can be fulfilled only by those who are educated both in the new informational skills, and also classically and historically, which is to say, it can only be fulfilled by an elite which preserves book knowledge, and thereby also, even in the computer age, a modern historical consciousness.

CHAPTER 8

ON ESPERANTO: USAGE AND
CONTRIVANCE IN LANGUAGE

In an oft-quoted passage of his intellectual autobiography, Rudolf Carnap recounts how vehemently negative Wittgenstein became when, at their very first meeting in 1927, the latter was told about Carnap's interest in "the problem of an international language like Esperanto". "A language which has not 'grown organically'", writes Carnap, seemed to Wittgenstein "not only useless but despicable".[1] So: Was Wittgenstein's aversion justified?

The adjective 'organic', just like the related term 'natural', admits of degrees; naturalness is a matter of aspects, of levels. And the first observation I would like to make is that in its more basic aspects the natural or "organic" development of language seems to lead to a kind of *fragmentation*. As the linguist Charles Hockett writes, a language is a system in which every actual utterance at one and the same time both conforms to the system *and* changes the system, however slightly.[2] If a language is spoken over a large area, the dynamics of change must lead to differentiation. This means above all fragmentation by sound-change, which is only in part counterbalanced by phonetic, syntactic, and semantic borrowings. On another level, however, there are political and social processes such as territorial conquests or the development of commerce, which *may* lead to strong convergences. And although these are not brought about by the inner dynamics of speech or writing, they might still count as natural, or spontaneous, from a linguistic point of view.

What, now, of the *conscious* or deliberate enforcement of linguistic convergence, i.e. *language standardization*? Here the major factor is of course the emergence of *writing*, which is "the medium of communication between speakers separated in time and space".[3] And it seems to me that it is a hitherto neglected insight that one basic reason for Wittgenstein's hostility towards meddling with language was that he had a particular sensitivity to the *sounded word*, to the specifically *oral* traits in the culture surrounding him, both the pre-literate culture of everyday language use, and indeed the post-literate culture of radio and cinema.

I have for some years maintained that Wittgenstein was a traditionalist.[4] And it seemed to me to be rather obvious that it was this traditionalism which explained Wittgenstein's repulsion at the idea of an artificial language. Now, however, I have come to believe that it was Wittgenstein's sensitivity to the oral aspects central to any and every culture which here plays the primary role. This sensitivity to oral communication was behind both his traditionalism[5] *and* his hostility to Esperanto,[6] as when Wittgenstein talks of "the feeling of disgust we get if we utter an *invented* word with invented derivative syllables. The word is cold, lacking in associations, and yet it plays at being 'language'. A system of purely written signs would not disgust us so much".[7] To be sure, one can identify certain modes of discourse which promote linguistic rigidity or formality even in non-literate societies. However, it is characteristically with respect to *written language* that rigid standards become established; and it is the spuriousness of just such standards which is often alleged by linguists. Thus Robert A. Hall writes:

'Correct' spelling, that is, obedience to the rules of English spelling as grammarians and dictionary-makers set them up, has come to be a major shibboleth in our society. This shibboleth serves, as does that of 'correct' speech, as a means of social discrimination...[8]

It is interesting to observe what little weight such a statement actually carries. Hall himself, as it becomes obvious in the earlier stages of his argument, is no enemy of standardized spelling; but he would want the standards to be *rational*. Thus he writes:

In reality, it does not matter what letters are used to represent one sound or the other; provided the letters are used consistently, they give a good representation, and if they are used inconsistently (as in English spelling) they give a poor representation. In most civilizations, the spelling of the language has come to be fixed through custom, and becomes something to which people have an unreasoning emotional attachment, despite any shortcomings it may have. Such a fixed customary spelling as we have for English, French, German, etc., is usually called *conventional spelling* or *traditional orthography*.[9]

Hall refers to languages which have "a simple, relatively accurate conventional spelling", like Italian, Hungarian, or Finnish,[10] and to the even more accurate ways of phonetic representation in so-called transcription systems.[11] He envies the Hungarians who have (as he mistakenly thinks) very few problems in the way of spelling; but what really fascinates him is the "radical and effective" dream of making English spelling *completely phonemic*. A mere dream though, for such a project is, as Hall puts it, a quite unattainable goal. The obstacles are not linguistic in nature, but political and economic. ... such a drastic reform could never be adopted by any other than authoritarian means... There are too many vested interests, also, standing in the way, quite aside from the widespread conservatism and inertia of the general public: just to mention one aspect of the matter, the trouble involved in reprinting all our books would be enormous, and the expense staggering. A great many important, but not popular, books might go unreprinted and fall into undeserved oblivion; for, within fifty years, everything that is now in print would be ... obsolete and unreadable... – And yet a partial reform would be useless; it would fall between two stools, that of rational and thorough improvement, and that of sticking to what we have, which at least has the virtue of familiarity and ease through long acquaintance and practice.[12]

There have been attempts to evaluate and to change languages along the dimensions of efficiency, rationality, clarity, economy, beauty, and the like. Reformers have been struck by the impression "that language lags behind thought" and were prepared to break with the past if that was the price of creating an "adequate expression for new meanings".[13] Alternatively, they were concerned to preserve what they considered to be the specific spirit of a given language – while changing the language itself. But none of these criteria or attitudes seems to cut any ice when put into practice. One of the earliest successful European language reforms has been the Hungarian reform at the beginning of the nineteenth century – "successful" in the sense of having in fact altered the language.[14] Describing this reform Géza Bárczi remarks that many newly contrived words of unhappy coinage, "running counter to the traditions of the language", soon became standard elements of popular vocabulary, some of them even entering the vernacular.[15] Similar observations have been repeatedly made. Of recent date are those of Yafa Alloni-Fainberg, on reforms pertaining to modern Hebrew. There are words, Alloni-Fainberg reports, which gain wide currency the moment they are invented, and nobody seems to know "how and why they spread as they do". On the other hand

there are words that though they have been conceived with great care, though much thought has been devoted to them, though they are based on good, indigenous roots and forms, yet they never catch on and nobody uses them naturally.[16]

If belief in the importance of "indigenous roots" thus turns out to be a romantic illusion, the idea, on the other hand, of language catching up with thought is of course a manifest absurdity. All a language might have to catch up with is *another language* – if, namely, a nation experiencing retarded development is striving to translate into its own vocabulary terms deriving from a foreign culture regarded as more advanced. And as to the criteria of efficiency, economy, etc., they certainly seem to be delusive. As Greenberg has put it, "it appears that natural languages are all very much on the same level as far as efficiency is concerned".[17] Einar Haugen reminds us of

the well-known tendency of languages to add complexity in one area when reducing it in another, for example, the substitution in many Indo-European languages of prepositions and fixed word order for disappearing case endings. Each living language appears to have achieved a form that is kept in equilibrium by the law of least effort on the part of its users.[18]

Analyzing empirical findings, Alloni-Fainberg regards as definitely refuted the hypothesis according to which "the length of a neologism" is of relevance for its acceptability.[19] On a more general level, C.A. Ferguson finds that evaluation of languages in terms of purity, beauty, and efficiency, is but a case of *rationalization*: the criterion of purity is inevitably capricious, aesthetic judgments reflect "not direct natural response to linguistic features but feelings of appropriateness due to customary use", whilst definitions of efficiency become feasible only if a specification of extra-linguistic goals is provided for.[20] Evaluations of language are, ultimately, *social evaluations*. "In all speech communities", Ferguson writes,

language users sometimes explicitly call attention to particular features of language structure or use as signals of group identity, disapproved behavior, objects of correction or other social values.[21]

To place high value on a particular language or on particular linguistic standards means, in other words, to accord high social prestige to the group using that language or conforming to those standards. This state of affairs has been widely recognized, but it is generally misrepresented. What it implies is that the epistemological significance of social elites must be underscored. Instead of this, however, it is taken to signify an essential vacuity of the distinction between what is correct and what is incorrect.[22]

In language as elsewhere, elites give rise to and sanctify order. The nature of the elites in question will of course vary greatly from case to case. They may be, typically, holders of political power. Thus the compilation of the first grammars and dictionaries of modern European languages in the fifteenth and sixteenth centuries coincided with the rise to wealth and influence of particular countries and groups.[23] Alternatively however the elites may consist of individuals whose intellectual prestige is exceptionally high. Thus the English accepted, as Haugen puts it, "the decrees of a private citizen, Samuel Johnson, whose dictionary (1755) became the first important arbiter of English".[24] Or they might embody a popular opposition against established elites, constituting, thereby, a counter-elite whose linguistic usage is built on rural dialects or everyday urban speech.[25]

It is, now, precisely because the emergence of relevant elites is hardly conceivable that the success of any attempt at what one could call the supreme linguistic contrivance, the establishing of a *universal language*, seems to be so very unlikely. At issue is not whether such languages can be constructed – a number of them have in fact been offered up for use. The question is whether they can, to any significant extent, fulfil their purpose. The artificial auxiliary language that has gained the relatively widest currency in the course of its development (now spanning over a hundred years) is Esperanto; but it has certainly not become a universal language, and the reasons for this failure are not difficult to see. Created by a Polish physician of Jewish origin and first presented in Russian, Esperanto has a rather simple grammatical system and a root lexicon based on Romance and, to a lesser degree, on Germanic languages. As far as the words of Romance origin go, the basic vocabulary of Esperanto is obviously related to their simplifying Russian/Cyrillic transcriptions, like when the French word *billet*, ticket, becomes *bilet* in Russian and *bileto* in Esperanto. The source of the German borrowings, on the other hand, is actually a simplified form of Yiddish, e.g. the German *gaffen*, to gape, is *gapen* in Yiddish, and *gapi* in Esperanto. The mechanisms of simplification and abbreviation so basically involved in its emergence made some linguists associate Esperanto with the Creole languages;[26] Hall associates it with Pidgin.[27] Wittgenstein somewhat misjudged the causes of his nausea when he picked on the element of *invention*.

At the first world congress of Esperantists in 1905 a stock vocabulary was accepted and declared as inviolable: this was the basis from which to form new words by derivation, and to which to add new words by borrowing, if necessary. However, the direction and extent of innovations became an issue which, since there was no generally accepted authority to which to appeal,[28] repeatedly led to tensions. Already during the lifetime of the initiator of the language, the Esperantists split into two groups, the "conservatives" adhering to the original form, and the "progressives" introducing a modified version called Ido. Within Esperanto proper, there arose the strife between "fundamentalists" and "neologists". But the essential problem is not that of intentional reform. The real difficulty is that, as soon as a language comes to be spoken by a greater number of people and in a great variety of situations, changes in usage inevitably crop up.

There are by now actually those who are native speakers of Esperanto, having been brought up by Esperantist parents and using, from their earliest childhood, Esperanto as a normal family language: within such families, however, spontaneous linguistic changes have inevitably taken place, a *family dialect* has emerged, and the changes in question have to be considered in official Esperanto.[29] On the other hand the process of differentiation is of course generally enhanced by the fact that the overwhelming majority of Esperanto users are native speakers of some other language. As Karl Brugmann foresaw it in 1907:

um ihren Zweck auf die Dauer erfüllen zu können, dürfte die internationale Sprache bei und nach ihrer Einführung keinen stärkeren Einwirkungen verfallen von seiten der verschiedenen Sprachen, neben denen sie von einem jeweils kleineren oder größeren Bruchteil der Bevölkerung zu gebrauchen wäre. Über die buntscheckige Sprachenkarte Europas ausgegossen, müßte die Hilfssprache ihre eigene Farbe auf die Dauer wahren können: es dürften keine Verfließungen eintreten, die zu solchen Verschiedenheiten führten, daß das gegenseitige Verstehen sehr wesentlich beeinträchtigt und allmählich ganz aufgehoben würde. Wie das abzuwenden wäre, sehe ich nicht.[30]

To put it in a nutshell: insofar as Esperanto becomes continuously more differentiated, it undergoes a development parallel to certain phases in the development of natural languages. At the same time however it gradually looses its original character, and ceases to be either simple or universal.

The term for "crocodile" in Esperanto is *krokodilo*. From this term the verb *krokodili* was derived at some Esperanto conference, meaning "Esperantists talking to each other in their mother tongue"; while *aligatori* has come to mean "talking neither in Esperanto, nor in the mother tongue", e.g. a German and a French talking to each other in English. These semantically anomalous verbs are something like *idioms* in a language which by its original construction excludes idiomaticity. Idiomaticity however seems to belong to the very essence of a natural language.[31] But to say that languages are essentially idiomatic is to maintain that learning a language means more than merely learning its rules and some basic vocabulary: it means memorizing a great amount of grammatically idiosyncratic forms. Although syntactic rules obviously fulfil important functions, "there is also much in linguistic activity which seems to be more plausibly described as the recall of quite specific memories".[32] This recognition of course runs counter to the view which for a time, in the wake of Chomsky, had been quite dominant: i.e. the view that everyday linguistic activity must be a fundamentally creative process, since a fluent speaker is able (or so it was thought) to produce and understand indefinitely many sentences "that are wholly novel to him. ... The striking fact about the use of language is the absence of repetition: almost every sentence uttered is uttered for the first time".[33] The Chomskyan idea of linguistic creativity as based on recursive syntactic rules – an idea never accepted in Wittgensteinian circles – has been shown to be untenable e.g. by Hockett in his arguments directed against the presupposition that language is a well-defined system;[34] and I side with Bolinger for whom language, far from brimming with creativity, displays, rather, a "deadly repetitiousness".[35] Bloomfield, significantly, found it an excellent method in language teaching to skip difficult grammatical explanations and have the pupils "memorize a dozen or so of typical sentences" instead.[36] This method was, of course, not particularly new. Among the earlier great linguists, e.g., Heyse argued in its favour:

Der praktische Sprachunterricht muß den natürlichen Weg der Sprachaneignung möglichst nachzuahmen und zu ersetzen suchen. Er muß mithin vorzüglich das Gedächtniß in Anspruch nehmen und durch mannigfaltige Uebungen ein Gefühl für das fremde Idiom zu wecken suchen. ... Die Reflexion, das Bewußtsein der Gründe und Gesetze, wirkt dabei leicht störend und hemmend, indem sie bedenklich macht. Daher die Erscheinung daß die gründlichsten Kenner einer Sprache dieselbe oft nicht so geläufig sprechen und schreiben wie Ungelehrte und Frauenzimmer.[37]

Indeed learning a language even involves the process of gaining a knowledge of the characteristic situations in which specific linguistic forms are applied, as this is of course implied in Wittgenstein's formula that meaning is use.[38] The various meanings of, say, the word 'bachelor' are learnt by encountering sentences like "He became a bachelor in 1965" and "The old bachelor finally got married" – sentences conveying semantic *and* factual information at the same time.[39] One does not understand the word 'bachelor' unless one has knowledge of a world of marriages, knights, universities, and fur seals; the knowledge of one's language cannot be separated from the knowledge of one's world. Semantic conventions are embedded in the inherited conventions that constitute a form of life,[40] speech habits cannot be "segregated ... from the totality of habits that constitutes the particular speaker's ... whole share of the lifeways of his community".[41] For Esperanto to be a full-fledged language it would have to be the language of a self-contained society. Or, to put it in another way: the idea of a universal language is as illusive as that of a universal form of life.

Now I am of course aware of the recent exchange between Newton Garver, Rudolf Haller, and Eike von Savigny, on the issue whether there is one single all-encompassing *Lebensform*, or at least an all-encompassing *allgemeine menschliche Handlungsweise*.[42] What I have to point out is that this is not the issue at stake here. Clearly, no one would ever want to invent a language in which only the anthropologically universal features of life are taken care of. As far as the dream of a universal language is concerned, the problem is, rather, whether one could imagine individual variations in life-styles being reduced to an insignificant level. If we take such a putative reduction to be an outcome of political processes, then the answer is of course that such reductions should not, and probably could not, ever take place. And if we take the reduction to be the spontaneous end-effect of technological processes – mass media, global communication, computer instructions, an ever more impoverished international English, and so on – then my answer is that I am not attracted by such developments at all, I don't think they are plausible, and at any rate any language which would ensue from such spontaneous processes would not be a *contrived* language.

CHAPTER 9

HEIDEGGER AND WITTGENSTEIN

The words with which Karl-Otto Apel introduced his paper "Wittgenstein und Heidegger" of 1967 – namely that the juxtaposition of these two names in a philosophical essay "still awakens some indignation"[1] – have lost nothing of their force in the succeeding decades. On the growing number of comparisons between the two philosophers which have appeared in connection with their common centennial in 1889, one could read in a prominent German weekly:

Whatever might be found or invented in the way of so-called parallels at airless heights of abstraction – an abstraction that is for this reason always "right" because it has nothing to say – a more absolute contrast of thought, of style, of total intellectual appearances can hardly be imagined than that between these two most influential philosophers of the 20th century.[2]

In opposition to this, however, I wish to defend a view advanced most recently by the literary theorist George Steiner. As Steiner maintains in the "Preface 1989" to his book on Heidegger, it is

the deep-seated connections between Wittgenstein and Heidegger, the two most outstanding, apparently so antithetical philosophical-linguistic thinkers of our time, which in the future will provide the most fruitful soil for research and understanding.[3]

In what follows I want first of all to summarize very briefly the most important parallels that have been put forward in the literature, to extend these parallels somewhat, and finally, as my main concern, to sketch a conceptual framework for their interpretation, a framework that is broader than is customary.

The all-embracing theme of Heidegger, and of Wittgenstein, *the overcoming of Western metaphysics*, lies at the centre of Apel's essay. Apel mentions the fact that both in Wittgenstein and in Heidegger it is properly necessary to distinguish two creative phases in their work – which is to say, not only the early Wittgenstein of the *Tractatus* and the later of the *Philosophical Investigations*, but also the Heidegger before and after the so-called *Kehre* or "turning". Yet he emphasizes also that this fact gives rise to no difficulties in connection with our present theme.

93

For

the setting at a distance of traditional metaphysics through a critical examination of
the question of meaning is ... the perspective which, in the form of a suspicion of
meaninglessness, connects the early to the later Wittgenstein, and on the other hand in
the case of Heidegger, too, defines the concept that remains constant in the change
from existential philosophy to the history of being.[4]

Apel notes a parallel between on the one hand the early Wittgenstein's
distinction between what can be said and what can only *show* itself, and
on the other hand the "ontic-ontological difference" of Heidegger's *Being
and Time*.[5] However, Heidegger already in *Being and Time* made the
attempt, "with the aid of the non-theoretical language of everyday life ...,
as it were to undercut the language of traditional onto-logic in order to
bring to expression a more original understanding of being". Wittgenstein, in contrast, attempted something similar only in his later phase,
which is to say, after 1929.[6] Apel holds that Wittgenstein's everyday
"language games", interwoven in the practices of life, unmistakably recall
Heidegger's "being-in-the-world". In the background of the idea of
language games there is, of course, a quite specific interpretation of
meaning: the air of affinity forces itself upon us, Apel writes:

when we try to read §§ 1–38 of Wittgenstein's *Philosophical Investigations* as it were
through Heidegger's eyes: One finds here, crudely put, a challeng to the thought-
model which has dominated the logic of language since Aristotle, according to which
the words of language have "meaning" through "designating" something, and this
means – if one follows the underlying paradigm to its sources – that words are
"names" of "existing things" or "objects".[7]

Indeed, as Carl Friedrich Gethmann points out, one can find elements of a
"semantic operationalism" in Heidegger already in 1921/22. One can
speak, in Heidegger's case, of an "enactment theory" of meaning parallel
to the "use theory" defended by Wittgenstein. This, according to Gethmann, amounts to the claim that

he knows the meaning of an expression who knows what must be brought about or
enacted if the object is to be had to which the expression is applied. That which is
primary is accordingly "the having of the object", the "mode of access". The
denotation of the expression is not a thing, but a "having".[8]

And thus in Heidegger as in the later Wittgenstein the refusal of the
received theory of meaning makes possible an overcoming of mentalism
– for one must no longer associate mental *objects* with mental terms.
There is according to Heidegger, Gethman holds, "no reasonable sense in

which we could distinguish between the inner and outer side of our relation to objects".[9] Apel, too, emphasizes that the common critique of language of Heidegger and Wittgenstein leads in each case as a matter of course to a philosophy of intersubjectivity, and thus here, too, to a rejection of the metaphysical heritage. To Heidegger's concept of "being-with-others" there corresponds Wittgenstein's argument that there could never be a so-called "private language"[10] – and this for *conceptual* reasons.

Apel points to similarities which he discovers between Heidegger's polemic against Descartes in *Being and Time* and the Descartes-critique – "in the spirit of Wittgenstein" – put forward by Gilbert Ryle in his *The Concept of Mind*.[11] Michael Murray followed up these similarities,[12] and the fact that Ryle had published a review of *Being and Time* in *Mind* in 1929 was of course never completely forgotten.[13] Murray however succeeds in presenting certain parallels between *Being and Time* and *The Concept of Mind* (1949) so penetratingly, that he can maintain a view of Ryle which sees the thinking of the latter as independent of Wittgenstein and as deriving, rather, from Heidegger. Now of course the first half of this thesis is not in every respect convincing. Murray leaves out of account the fact that transcriptions of the so-called *Blue and Brown Books* were rather freely available in England in the '30s and '40s, and that these texts already formed, without doubt, a chapter in the development of Wittgenstein's later philosophy. Even the suggestion, however, that Heidegger might have exerted a crucial influence on Ryle – who does after all count as a sort of poor man's Wittgenstein – is more than revealing.

Murray's work has been taken up in the German-speaking world above all by Thomas Rentsch in his *Heidegger und Wittgenstein* of 1985, wherein the Heidegger–Ryle background is exploited in order to emphasize certain Heideggerian anticipations of basic ideas of the *Philosophical Investigations*. "'To imagine a language", says Rentsch, citing Wittgenstein, "'means to imagine a form of life.'" And:

"Now what do the words of this language *signify*? – What is supposed to shew what they signify, if not the kind of use they have?" Later Wittgenstein compares the function of words to *tools* in a *tool-box*.

This "situational pragma-semantics", Rentsch writes,

can be found in essence 20 years earlier in §§17f. of *Being and Time*. The holistic aspect, the connection between reference and use and the tool-character are there clearly put forward.[14]

The concept of a tool points to a sense of knowledge as an *ability*, to *skills*, to the embeddedness of what is cognitive in the practical realm. The important and – as I wish to suggest – symptomatic work in which this dimension of the Heidegger–Wittgenstein relationship is treated, is Herbert Dreyfus' book on the limits of artificial intelligence.[15] The problem of the representation of knowledge *cannot be solved*, Dreyfus writes here appealing to Heidegger and Wittgenstein, because knowledge is wrongly conceived – in conformity with the Platonic-Aristotelian tradition – as something abstractly *representable*. Action does not rest as it were "on a practical syllogism";[16] rather, a sign has meaning only "'within-the-world in the whole equipment-context'";[17] there is a "background of practices which are the condition of the possibility of all rule-like activity".[18]

Now, however, I want to point to an additional parallel between Heidegger and Wittgenstein, one which has been less considered in the literature: a parallel in their world views. As to Wittgenstein, what I have to say here concerns his later writings which, even though not constituting a complete break with his earlier thinking, are still essentially different therefrom and for us today indubitably more important. The *Weltanschauung* by which not only Heidegger but also – and in an even more direct manner – Wittgenstein is marked, is a conservative one. More precisely it is one of traditionalistic conservatism, with a distinctive aversion to everything that has to do with the so-called "Enlightenment", with modern science, with modern industry and technology, and with contemporary modes of life in general. That such an attitude was indeed characteristic for Heidegger is something that can, I think, be taken here for granted. In regard to the dimension of traditionalism in Heidegger's thinking it should suffice to recall the resounding words of the *Spiegel*-interview: "everything essential and great has come about only because man has a homeland and is rooted in a tradition".[19] And let me mention the essay "Hebel – der Hausfreund", in which Heidegger refers to dialect as the "secret source of every naturally-grown language" and speaks of everything high and valid as dying "as soon as a language must do without its influx from this source".[20]

As concerns Wittgenstein, now, one must point above all to the traditionalistic aspects of his concept of "form of life", as also to his emphasis, in some of the most decisive passages of the *Philosophical Investigations* (§§ 198 f.), on *habits, customs, institutions*, and of course also to the many relevant passages in the volume *On Certainty*.

This traditionalism is, in Wittgenstein even more than in the case of Heidegger, a part of a conservative conception of history. It is not by accident that Wittgenstein chose as the motto of the *Philosophical Investigations* Nestroy's remark as to the illusive character of the notion of progress. It is not by accident that in the foreword to this work we read of "the darkness of this time". The spirit which marks Wittgenstein's efforts is, as he himself wrote in 1930, "different from the one which informs the vast stream of European and American civilization in which all of us stand".[21]

Heidegger's attitude to modern technology is immediately connected, as is well known, with his interpretation of modern science; the latter, however, is in turn bound up with his view of the logical-metaphysical philosophy of the West since Plato and Aristotle as an aberration of thought. Modern technology is rooted in a "forgettenness of being" which set in already after Parmenides.

A similar, if less articulated, argument can be detected, now, in Wittgenstein. The "apocalytic view of the world", Wittgenstein writes in 1947,

is that things do *not* repeat themselves. It isn't absurd, e.g., to believe that the age of science and technology is the beginning of the end for humanity; that the idea of great progress is a delusion, along with the idea that the truth will ultimately be known; that there is nothing good or desirable about scientific knowledge and that mankind, in seeking it, is falling into a trap. It is by no means obvious that this is not how things are.[22]

Modern humanity – as one might express Wittgenstein's view (and here it is the traditionalist-nostalgic side of his conservatism which predominates) – has fallen into a *false form of life*. But from a false form of life – and this is a seldom recognized argument of Wittgenstein's – there grows a false philosophy. Thus consider what Wittgenstein has to say in the unpublished *Big Typescript*, compiled in 1933:

Human beings are deeply embedded in philosophical – i.e. grammatical – confusions. Freeing them from these *presupposes* tearing them away from the enormous number of connecting links that hold them fast. A sort of rearrangement of the whole of their language is needed. – But of course that language has developed the way it has because some human beings felt – and still feel – inclined to think that way. So the tearing away will succeed only with those in whose life there already is an instinctive revolt against the language in question and not with those whose whole instinct is for life in the very herd which created that language as its proper expression.[23]

What stands out here is the suggestion that the possibility does after all obtain of rescuing humanity from its false mode of life and thereby also from false language.[24] This possibility, however, is a slight one. "Science and industry, and their progress", Wittgenstein writes,

might turn out to be the most enduring thing in the world of today. Perhaps any speculation about a coming collapse of science and industry is, for the present and for a *long* time to come, nothing but a dream; perhaps science and industry, having caused infinite misery in the process, will unite the world – I mean condense it into a *single* unit, though one in which peace is the last thing that will find a home. – Because science and industry do decide wars, or so it seems.[25]

But it could also be that humanity is rescued precisely through science – as it were in a process of self-destruction. As Wittgenstein notes in 1946:

The hysterical fear over the atom bomb now being experienced, or at any rate expressed, by the public almost suggests that at last something really salutary has been invented. The fright at least gives the impression of a really effective bitter medicine. I can't help thinking: if this didn't have something good about it the *philistines* wouldn't be making an outcry. But perhaps this too is a childish idea. Because really all I can mean is that the bomb offers a prospect of the end, the destruction, of an evil, – our disgusting soapy water science. And certainly that's not an unpleasant thought; but who can say what would come *after* this destruction? The people now making speeches against producing the bomb are undoubtedly the *scum* of the intellectuals, but even that does not prove beyond question that what they abominate is to be welcomed.[26]

In Wittgenstein as in Heidegger, therefore, the atomic bomb figures merely as the sign of a development which would amount to a decline even without this consequence. It is, as Heidegger writes, not that man must "live through atomic energy; at best he must perish, which is to say lose his nature ... and this even if atomic energy is used only for peaceful purposes".[27] And in Heidegger, too, we find the hope – not easy to grasp conceptually – that "the coming to presence of technology harbors in itself ... the possible arising of the saving power".[28]

Perhaps I have now at least hinted at the way in which Heidegger and Wittgenstein, in spite of all apparent opposition in their style and approach, are related to each other through similarities which seem to be neither superficial nor merely occasional. But one would surely better be able to cope with the relationship between Heidegger and Wittgenstein if one had an *explanation* for these similarities, if, that is to say, one could give *reasons* for them. And the simplest explanation would of course be one able to establish *influences* between the two. Heidegger, now, is known to have mentioned Wittgenstein in the winter semester 1966/67;[29] one cannot, however, assign much significance to this late occurrence. It is reported also that Heidegger had some years earlier referred to Wittgenstein as a "crass positivist".[30] Wittgenstein on the other hand expressed some understanding for Heidegger's remarks towards the end of 1929 about being and *Angst.*[31] This reference seems to relate to the lecture "What Is Metaphysics?" an echo of which can still be heard in the "Lecture on Ethics" delivered by Wittgenstein in 1930.[32] The hypothesis of Thomas Rentsch according to which Wittgenstein's later philosophy developed under the influence of *Being and Time*, though fascinating, seems to be without any support. What seems to me to be crucial, in all of this, is the fact that in Wittgenstein's notebooks from 1929-1931 – and these were the decisive years – the theme of Heidegger is not mentioned. Wittgenstein's later philosophy takes its origin, quite clearly, from an argument with Plato, with Frege's Platonism, and with his *own* earlier Platonism.

This last remark provides an answer also to the question whether there may not be common intellectual roots in which we might search for the reasons for the affinities which here concern us. Austrian philosophy certainly plays something of a role here: in this connection it is significant that Heidegger should have preferred in the 1920s the *Logical Investigations* of Husserl rather than the later works of the master. And one can of course refer here also to the influence of Dostoevsky and Spengler on both Heidegger and Wittgenstein.[33] Yet explanations of this sort do not suffice for the solution of our present problem; nor, indeed, do the usual references to the intellectual consequences of World War I.

The decisive element in Wittgenstein's later philosophy is his conception of meaning. By conceiving meaning in terms of *use* and not in terms of one or other *referent*, he manages to avoid the ontology of abstract objects and the epistemology of mental entities. And because he is not forced by epistemological reasons to accept talk of interiority, he can escape philosophical individualism and thus stand in opposition to the modern age.

Now the use theory of meaning lets language appear as an *activity*, as something *living*. Wittgenstein's question: "Every sign *by itself* seems dead. *What* gives it life?"[34] – does not arise in relation to spoken, but only in relation to written language.[35] The use theory of meaning thus stands under the impact of orality. Indeed Wittgenstein himself seems to have had a difficult relationship to written language, in particular to written language in its most highly developed form: the printed book. I have in mind his poor orthography; his anachronistic predilection for having people read texts to him out loud; the observation often expressed that his favourite readings he really knew by heart; his inability or unwillingness to put together what one would call a *book* or *treatise* in the modern sense; and of course his reluctance to *publish* his writings. For *spoken* language, in contrast, he had a quite peculiar empathy, and an important pointer here might be his emphasis on the vernacular. His *Wörterbuch für Volksschulen*, compiled in the early 1920s in the course of his activity as an elementary school teacher in Lower Austria, deliberately reflects the pupils' own linguistic usage; it does not avoid dialectal expressions, and includes some very common words which are typically used in *speech*, like "geh!" or "hierher". It even utilizes dialectal *pronunciation* in order to bring home some grammatical points.[36]

From a historical point of view the transition from orality to literacy occurred in Greece, and Plato was the philosopher whose work first bears the marks of an entirely interiorized experience of language as *written*, *seen* language, rather than as something heard[37] – a circumstance made all the more conspicuous by his own expressed aversion to the technology of writing. It is Plato's experience of the written language which forms the background of those of his arguments which became so central to subsequent ontology; and it is these arguments which Wittgenstein over and over again faces in those remarks in which he arrives at the theory of meaning as use. The point to which Wittgenstein wanted to return is the point at which Plato had taken the wrong turning. It is a striking fact that Wittgenstein, whose comments on earlier figures in the history of philosophy are notoriously scarce, often mentions, and indeed quotes, Plato.[38] In the 1930–31 notebooks, in particular, Wittgenstein repeatedly refers to passages in Plato where the connections between a misguided theory of meaning on the one hand, and an extravagant ontology on the other, are especially conspicuous. Thus for example the important entry of July 9, 1931 – "To understand the meaning of a word means to be acquainted with, to understand, a use"[39] – is followed, on the next page, by a quote from the *Cratylus* on the assumed, or desirable, likeness between signs and the objects they stand for.[40] And on the following page we find a quote from the *Theaetetus* on the peculiar state of affairs that anything represented must be, somehow, real.[41] The latter elicits a comment from Wittgenstein: "How infinitely simple this problem is! And how strange that it could have been regarded as a problem at all",[42] followed, on the next day, by the remark: "I do not find in Plato the preliminary answer to a question like 'what is knowledge': Let us try and see how this word is used".[43] It is as if Wittgenstein were conscious of the fact that he has, in philosophy, to start anew at the very juncture where Plato's arguments went astray. "What marvellous people we are", he puts it in 1931, "to have solved these ancient problems! – No, time has changed us & the problems have vanished".[44]

We are back once more, it is clear, with Heideggerian themes. Heidegger's efforts to return to Greek thought as it was before Plato are, after all, one of the main characteristics of his work. And that his philosophy, both in its terminology and its arguments, is put forward in the spirit of orality – that he conceives of language as *speech* – should be immediately evident to any reader of his writings. This much should, however, be said: that one of the main concepts of his later thinking – the concept of *Sage* (meaning both "saga" and "saying") – constitutes a reference that is from the historical point of view entirely coherent. I adopt here a consciously demythologizing standpoint and direct myself against, say, the interpretation of von Herrmann, according to which "the speaking of speech is the saying of the saga. The term 'saga' is formed by the substantive of the verb 'say' and has nothing to do with what we normally call 'saga'".[45] If Heidegger tells us that "myth" means: "the telling [sagende] word", and that the Greeks experienced their "lofty poetry" "thoughtfully";[46] if he connects the saga with "the song that says by singing", talks of the saga as in need of (*braucht*) sounding words,[47] and finally when he writes: "The union of mortals releases the human being into custom" (*Brauch*)[48] – then I would like to conceive these passages in their straightforward sense, namely as references to an era when the memory of society was preserved in handed-down sayings, sagas, in oral tradition, where poetry was the custodian of knowledge, and where custom and habit were the instruments by which the polity was organized.

The spoken, resounding, heard language constitutes the primary environment of the individual human being. In *Time and Being* Heidegger alludes to this state of affairs by the formula "*The existential-ontological foundation of language is discourse or talk*"[49] – where "discourse or talk". *Rede*, should quite unequivocally be construed in the dimensions of hearing,[50] of intonation and of modulation, and of course of silence.[51]

Now Heidegger does not want us to think of sound as of a *mediator* between separate individuals,[52] i.e. as of the bearer of meanings which as it were could be dissociated from the word[53] – however attractive an idea this might seem from the perspective of writing.[54] This, he holds, is the point at which the Greeks first went astray – and not just the Sophists and Plato, as he initially tended to believe, but already the very first Greek thinkers: "Nowhere do we find a trace of the Greeks' having thought the essence of language directly from the essence of Being. Instead, language came to be represented – indeed first of all with the Greeks – as vocalization, φωνή, as sound and voice, hence phonetically. The Greek word that corresponds to our word 'language' is γλῶσσα, 'tongue'. Language is φωνή σημαντική, a vocalization which signifies something."[55]

Heidegger's own conception, in contrast, is expressed by the formula "We have heard [*gehört*] when we *belong to* [*gehören*] the matter addressed",[56] which on the one hand of course corresponds to his basic philosophical tendency to construe Being as a unified whole of which we are a part and the separation from which as it were opens up the abyss of death – but which on the other hand is also an expression of conservatism, here coming to the fore in Heidegger's repeated references to the time-honoured relationships between "hear", "hearken", "to pay heed", "be in bondage to someone".[57] But since Heidegger – more clearly than Wittgenstein – realizes how limited is a conservatism if it is just a longing for the past, a "mere negating of the age",[58] he lets the profane-historical interpretation of these relationships recede, as it were, behind an interpretation in terms of the history of being. "For man", he writes, "becomes truly free only insofar as he belongs to the realm of destining and so becomes one who listens and hears [*Hörender*], and not one who is simply constrained to obey [*Höriger*]."[59] A similar method is followed with respect to the notion of "custom", *Brauch*. Heidegger's statement that man belongs to a custom which takes possession of him[60] could be a historical-sociological one; and his translation of Anaximander's "τὸ χρεών" as *custom* or *usage*[61] could indeed be accepted, having regard to the connotation of *social necessity*. Heidegger however avoids the historical interpretation, and chooses connotations pertaining to the history of *being*. Thus his fundamental recognition that Western philosophy, since its turning point in Plato, focusses on *seeing*[62] – that "in the age of the Greeks the world cannot become a picture", where for Plato it had to do so[63] – is not brought into connection with the novel epistemic situation arising with the shift of the educated classes from language heard to language seen: with the shift, that is, to language *written*. Still, Heidegger was somehow aware of this connection. His *Einführung in die Metaphysik* contains a conclusive passage to the effect that the Greeks "in a certain broad sense regarded their language optically, namely from the vantage-point of writing. There what is spoken comes to a standstill. Language is, i.e. stands under the authority of, the word as it looks, the written sign, letters, γράμματα. ... So, then, the science of language [*Sprachlehre*] down to our age is grammatically conceived."[64]

Heidegger and Wittgenstein worked out their philosophies simultaneously, in a life-world in which the spoken word – through radio[65] and film[66] – was once more making its presence felt in the academic world and gradually acquiring a position of dominance.[67] They both abhorred this world of chatter – just as Plato had in his day abhorred the superficiality of writing. Both are, in my view, the philosophers of a new orality – dreaming of pre-literal times, the times of the old orality – just as Plato had been the philosopher of literacy. That those working in the field of artificial intelligence research today appeal more and more to Wittgenstein, and even more insistently to Heidegger,[68] I find not at all perverse – rather, as I indicated earlier, symptomatic. Heidegger and Wittgenstein are not only the philosophical critics of the new technological age.[69] They are also – recall the word *saving* – its prophets.

CHAPTER 10

WRITING AND THE PRIVATE LANGUAGE ARGUMENT

The phrase "private language argument" refers to a philosophical strategy crucially informing Wittgenstein's later work. But while my aim here is an evaluation, from a specific point of view, of this very strategy, it should be stressed that the issue addressed by the private language argument is of course a more general one, occupying a central place in the paradigm defined by Descartes and Locke.

According to the approach I take here as my point of departure, mentalism, i.e. the philosophy of an "inner world", of inner mental objects and events – and, even more directly, philosophical *Platonism* – is intimately bound up with the *visual* experience of language, i.e. with the experience of reading and writing. This approach is influenced by authors whose ideas in this connection Wittgenstein too would have been in a position to pick up – even if he, apparently, did *not* do so. That the emergence of writing might possess a specific epistemological significance had been already sensed by Plato; this insight was then, in our era, rediscovered by Oswald Spengler. And there is an essay by Bronislaw Malinowski that merits particular interest in this connection, included as an appendix in the Ogden and Richards volume *The Meaning of Meaning*.[1] Here "primitive living tongue, existing only in actual utterance" is contrasted with "dead, inscribed languages". The former, Malinowski stresses, is "to be regarded as a *mode of action*, rather than as a *countersign of thought*".[2] In a primitive language, he writes, "the meaning of any single word is to a very high degree dependent on its context"; indeed it is dependent, as he puts it, on the *context of situation* – i.e., on the extra-linguistic environment. Written documents, by contrast, are "naturally isolated", the statements contained in them "are set down with the purpose of being self-contained and self-explanatory".[3] Spoken linguistic material "lives only in winged words, passing from man to man", word-meanings being "inextricably mixed up with, and dependent upon, the course of the activity in which the utterances are embedded".[4] Language in a preliterate culture, Malinowski emphasizes, is never 'a mere mirror of reflected thought". In *writing* however "language becomes a condensed piece of reflection". the reader "reasons, reflects, remembers,

105

imagines".[5] And it is significant that in Malinowski's estimate such reflection is a philosophically dangerous enterprise, leading to a "misuse of words", bestowing "real existence" upon meanings – giving rise, that is, to Plato's *ideas* and to medieval realism.[6]

Wittgenstein shared Malinowski's bias in favour of the spoken word. In Wittgenstein's case this bias was not, of course, based on anthropological field-work. Rather, I think, it had to do with his problematic relation to written language. I think it can be demonstrated that the drive behind Wittgenstein's private language argument is precisely his predisposition in favour of language spoken, as opposed to written. It has to be pointed out, however, that Wittgenstein was obviously not aware of this predisposition, and was indeed oblivious to the philosophical significance of the speaking/writing contrast. And I would say that much of the seeming depth and manifest obscurity of Wittgenstein's texts stems precisely from his *not* making an explicit and appropriate distinction between the oral-aural and the visual modes of communication. Nor has this distinction ever been drawn in the commentaries dealing with Wittgenstein's private language argument – the result being that neither those favouring interpretations emphasizing the individual speaker's essential dependence on his linguistic community, nor those inclined towards explications underlining the possibility of solitary rule-following, are able to recognize the validity of their respective opponents' positions.

Wittgenstein introduces the notion of a private language in § 243 of *Philosophical Investigations*, Part I, from which I quote at some length:

A human being can encourage himself, give himself orders, obey, blame and punish himself; he can ask himself a question and answer it. We could even imagine human beings who spoke only in monologue; who accompanied their activities by talking to themselves. – An explorer who watched them and listened to their talk might succeed in translating their language into ours. (This would enable him to predict these people's actions correctly, for he also hears them making resolutions and decisions.)

But could we also imagine a language in which a person could write down or give vocal expression to his inner experiences – his feelings, moods, and the rest – for his private use? – Well, can't we do so in our ordinary language? – But that is not what I mean. The individual words of this language are to refer to what can only be known to the person speaking; to his immediate private sensations. So another person cannot understand the language.

This is supplemented by the definition in § 269:

And sounds which no one else understands but which I "*appear to understand*" might be called a "private language".

However, the notion of such a language is already foreshadowed by the argument given in § 202:[7]

> "obeying a rule" is a practice. And to *think* one is obeying a rule is not to obey a rule. Hence it is not possible to obey a rule "privately"; otherwise thinking one was obeying a rule would be the same thing as obeying it.

Wittgenstein's thesis of course is that a private language is impossible. This thesis however, like the notion of a private language itself, is far from being unambigious, as is shown by the existence of innumerable commentaries on the relevant texts. The spectrum of interpretations extends from the classical one of Rhees, according to which the intelligibility of anyone's language is actually dependent on there being a community of speakers who use that language,[8] to Baker's and Hacker's argument, aimed at showing that "someone's mastery of a technique *need not* be related to regularity in the behaviour of *others*; that is relevant only to establishing mastery of a *shared* technique". It is wrong, they maintain, to take "Wittgenstein's conception of a practice to be necessarily a *social* practice". What Wittgenstein has proven, Baker and Hacker write, is that there can be "no such thing as one person's exercising a technique which nobody else *could* be taught to come to master".[9]

Now all these commentaries, to which one should add attempts at refutations like those of Strawson or of Castañeda, and partial re-interpretations like for example that of the Hintikkas, really address one central issue: the issue of *memory*, namely the question of how constancy and uniformity in the use of signs can be maintained, how *criteria of sameness*, ensuring regularity in the application of rules, can be established. Accordingly, the arguments, whether affirmative or critical, all allot a central place to *Philosophical Investigations* § 258:[10]

> Let us imagine the following case: I want to keep a diary about the recurrence of a certain sensation. To this end I associate it with the sign "E" and write this sign in a calender for every day on which I have the sensation. – I will remark first of all that a definition of the sign cannot be formulated [sich nicht aussprechen läßt]. – But still I can give myself a kind of ostensive definition! – How? Can I point to the sensation? Not in the ordinary sense. But I speak or write the sign down, and at the same time I concentrate my attention on the sensation – and so, as it were, point to it inwardly. – But what is this ceremony for? for that is all it seems to be! A definition surely serves to establish the meaning of a sign. – Well, that is done precisely by the concentration of my attention; for in this way I impress on myself the connection between the sign and the sensation. But "I impress it on myself" can only mean: this process brings it about that I remember the connection *right* in the future. But in the present case I have no criterion of correctness. One would like to say: whatever is going to seem right to

me is right. And that only means that here one can't talk about "right".[11]

The gist of what I will be trying to say here is that individual memory functions in an entirely different manner, and has a radically different scope, under conditions of an oral culture on the one hand, and under those of fully developed literacy on the other. The individual in an oral culture, even if helped by some rudimentary mnemotechnical devices, has no means to check the reliability of his necessarily fleeting recollections other than through the reactions of his fellow speakers. The literate person in contrast is essentially capable of comparing visual signs in order to establish the identity or difference of written texts.[12] This contrast is systematically obscured by Wittgenstein. His remarks contain references both to spoken and to written language; but his arguments invariably mobilize and exploit oral intuitions[13] – and the commentators follow spellbound. It would be as easy as it would be boring to show that wherever in the literature on the *Philosophical Investigations* the possibility of a private linguistic practice is entertained in an affirmative spirit, there crop up examples pertaining to written language, albeit never explicitly recognized as such, while the arguments directed against the intelligibility of private criteria, even when mentioning visual signs, actually play on the difficulty of comparing evanescent aural impressions separated in time from each other. Two examples will suffice: at a crucial juncture in their argument, Baker and Hacker of course find it necessary to refer to an imagined Crusoe's ability *to keep records*, or to employ, in his rule-formulations, visible and permanent *geometrical patterns*.[14] Or, on the other side of the fence, Kripke, when embarking upon his exposition of the Wittgensteinian "sceptical paradox", typically discusses addition between *small numbers* and the application of terms in *oral discourse*[15] – reference to adding on paper appears only later in the argument, when the normal reactions of the reader have been already stifled.

Let me dwell in somewhat greater detail on the various techniques of communication one might distinguish when attempting to answer the question whether languages private in Wittgenstein's sense are conceivable. I will distinguish between (A) talking-to-oneself, (B) "inner speech", (C) the spoken language of a strictly oral community, and (D) the spoken or written language of a literate – in the sense of *well-read* – individual.

Talking-to-oneself is a familiar – and, one would assume, philosophically trivial – phenomenon. It is, in Wilfrid Sellars' words, "the verbal imagery" which enables us to "'hear ourselves think'".[16] It is, as Strawson puts it, what occurs when one is "saying something to oneself in imagination", it consists of words that are "going through my mind".[17] What I here mean by talking-to-oneself is not merely, and not even primarily, what we indulge in when actually and loudly conducting a dialogue with ourselves; still, I do not assume that talking-to-oneself has a structure different from that of audible speech, or that it would give rise to epistemological issues different from those pertaining to the latter.[18]

Yet Wittgenstein discerns problems already at the level of talking-to-oneself. Before citing him on this however, let me make an observation: namely that, insofar as spoken language and written language have different structures, the talking-to-oneself of a highly literate individual typically has the structure of writing-down-visibly, rather than that of speaking-out-loud. The speech of the highly literate individual is, obviously, different from that of the uneducated, the illiterate, or indeed from that of the member of an oral culture. All the same, when *speaking*, the highly literate individual, too, performs speech acts, partakes in dialogue, uses elliptic constructions, swears and mumbles. However, when talking-to-*himself*, he typically becomes more elaborate; he thinks as if he were writing.

Now the point Wittgenstein will make here is that one's awareness of what one said-to-oneself is not, as it were, brought about by introspection. "That what someone else says to himself is hidden from me", Wittgenstein writes, "is part of the *concept* 'saying inwardly'. Only 'hidden' is the wrong word here; for if it is hidden from me, it ought to be apparent to him, *he* would have to *know* it. But he does not 'know' it; only, the doubt which exists for me does not exist for him".[19] I think Strawson is right when he detects here a *prejudice* against "'the inner'",[20] arguing that one can just as well *report*, and in this sense *know*, what words were going through one's mind, as one can repeat what one *said*. The "difference between *these* two cases", writes Strawson, is merely that

there is, and can be, no check (except the general reliability of my short-term memory) on my report of what words were going through my mind, whereas there can very well be a check on the correctness of my repetition of my audibly spoken words.[21]

Earlier in the essay here referred to Strawson discusses Wittgenstein's "E"-example – the assumed impossibility of recording, *in writing*, private sensations – and underlines that the "making of the marks" – namely by the keeper of the diary – "would help to impress the occurrence on his memory".[22] We have here, again, a notable instance of a commentator's making use, but not actually being aware, of the particular role played by the idea of writing in the private language argument. This much, then, about talking-to-oneself.

One likely candidate for the role of a private language of sorts is *inner speech*, in the interpretation the Russian psychologist Lev Semenovich Vygotsky gave to this term. In his book *Thought and Language*, published posthumously in 1934 and for some time now available in an English translation, Vygotsky clearly distinguishes inner speech both from "verbal memory" and from "subvocal speech",[23] stressing that the former has "special characteristics" due to its special function: "Inner speech is speech for oneself; external speech is for others".[24] Inner speech, in Vygotsky's analysis, originates in what Piaget called "egocentric speech"; but to see this, Vygotsky says, one first has to re-interpret Piaget. Whereas in Piaget's conception egocentric speech, as Vygotsky puts it, "has no function in the child's realistic thinking or activity", and "disappears together with the child's egocentrism", according to Vygotsky it is "a phenomenon of the transition from interpsychic to intrapsychic functioning, i.e., from the social, collective activity of the child to his more individualized activity".[25] Egocentric speech becomes more and more *inscrutable* as the child grows older and reaches school age;[26] inner speech is egocentric speech stripped of its "vocal aspect".[27] As Vygotsky writes:

With the progressive isolation of speech for oneself, its vocalization becomes unnecessary and meaningless and, because of its growing structural peculiarities, also impossible. Speech for oneself cannot find expression in external speech. The more independent and autonomous egocentric speech becomes, the poorer it grows in its external manifestations.[28]

Now the significance of Vygotsky's analysis for our present topic resides in the fact that he specifically deals with *writing*. Written speech is, he says, "a separate linguistic function".[29] Moreover, although he sharply contrasts the syntax of written language with that of inner speech – the latter is truncated beyond public comprehensibility, the former is "the most elaborate form of speech"[30] – still, he senses an intimate connection, sometimes indeed bordering on a curious identity, between inner speech and writing. Writing, Vygotsky emphasizes, "demands detachment from the actual situation", it "requires deliberate analytical action on the part of the child". Written language, Vygotsky says,

demands conscious work because its relationship to inner speech is different from that of oral speech: The latter precedes inner speech in the course of development, while written speech follows inner speech and presupposes its existence (the act of writing implying a translation from inner speech).[31]

We should notice that Vygotsky here actually speaks about *the translation of the inscrutable* – with the implication that it is *not* inscrutable for the thinker/writer, and even with the possible implication that written language, too, possesses a kind of inscrutability, namely for its readers. There is a certain "preponderance of the *sense* of a word over its *meaning*". Vygotsky says, which stems from the effect of linguistic context on the single word, and which Vygotsky discovers to be a characteristic feature of inner speech – illustrating it, surprisingly, by bringing in the issue of *literature*: "A word", he writes, "derives its sense from the sentence, which in turn gets its sense from the paragraph, the paragraph from the book, the book from all the works of the author".[32]

I am now – very briefly, since the ground is, I hope, prepared – coming to the issue of communication in primarily oral communities. Let me summarize what I wish to convey here by saying that, with one important qualification, spoken language without the background of literacy is the type of language in relation to which Wittgenstein's view that meaning is use seems to be correct; this is the language in which the use of words is in fact guided by public criteria; it is here that Wittgenstein's arguments against the possibility of a private language are truly convincing. The qualification relates to the role of *formulaic* – poetic, ritual – language.[33] Formulaic language clearly does not open up a space for self-referential analysis or critical discourse, but it does allow – once safely memorized – for a measure of cognitive distance from what is heard or said here and now.[34] Otherwise however the member of an oral society lives in an overwhelming, continuous cognitive noise, to which he listens and to which he contributes; he *echoes* the words of others, rather than "reflecting" upon them, "explaining" them, basing "insights" on them – note the dependence of these latter terms on visuality, i.e. on *written* language.[35] In an oral culture, *correct recollection* is a matter of consensus, memory is a collective faculty,[36] privacy of thought, in any philosophically interesting sense, inconceivable. Wittgenstein's relevant arguments, then, represent an epistemological predilection for a certain type of culture, for a certain form of life. This is pointed out in the book by Merrill and Jaakko Hintikka, where Wittgenstein is characterized as a "closet Tolstoyan in his philosophy of language".[37] But perhaps the expression "Tolstoyan" will not quite cover the attitude at issue here. I prefer, as I have always done, the term *conservatism*,[38] and suggest that Wittgenstein's oral bias, his interest in primitive societies, his authoritarian views on education, his traditionalism, his use-theory of meaning, and his arguments against the possibility of a private language together make up a coherent outlook.

Clearly, however, it is an outlook of only relative validity. For under conditions of a literate culture its very application is questionable, and this is especially so in the culture of the printed book. First, because the use-theory of meaning there loses its plausibility. Writing and reading are more aptly described in categories like deliberation and contemplation than in the terminology of actions. To paraphrase *Philosophical Investigations*, § 432: Every *written* linguistic sign is by itself *dead*. Its *meaning* gives it life. But what *is* its meaning? The question, apparently, will not be resolved by simple theories like the use- or the name-theories of meaning, or by dichotomies like that of sense and reference. It is however clear that in writing words do indeed *stand* for something: namely for *spoken* words; and, more interestingly, for *other words* within written language.

Secondly, Wittgensteins's outlook becomes questionable, when one realizes that in a literate culture *mentalism*, far from being a storehouse of illusions, provides the natural description of the writer/reader's relation to the written text. "Ideas" become perfectly permissible theoretical constructs on the basis of the observation that written words are objects of contemplation, of analysis, of comparison one with another. Where under conditions of orality the intentions of the speaker are inextricably fused with the surrounding situation, *written* words can meaningfully be regarded as separate targets of intentionality.[39]

Thirdly, once mentalist terminology is admitted, it becomes evident that *private* mental objects do indeed exist. And the terms by which these objects are named belong, as Castañeda would say, to a private *sublanguage*.[40] I will now introduce such a term: *Gregor Samsa* – partly an allusion to Wittgenstein's private beetle in § 293 of *Philosophical Investigations*. The definition is: "'Gregor Samsa' designates a literary figure with whose fate I have become acquainted in the course of reading Kafka".

The simple device which makes the object under consideration here private – namely that it is an object *I* had to become acquainted with, and thus your Gregor Samsa is not identical with my Gregor Samsa – is not in the least artificial. It mirrors, very precisely, a cultural condition under which personal knowledge is acquired in an individual way, namely by individual reading.[41] The definition would not work for an oral society, trivially because it mentions "reading", but more importantly because the abstract knowledge such a society has is all of a piece, knowledge is collectively acquired and uniformly possessed. *Literate* societies, by contrast, offer a measure of *cognitive freedom* to their members, freedom from handed-down and collectively held views.[42]

The assumption of cognitive autonomy however is tantamount to the assumption that under conditions of fully developed literacy the private language argument is *false*. Let me, by way of conclusion, try to indicate why this is so. Although at first sight it would seem that the impossibility of a private language does not preclude private opinions – we agree in meanings but as it were differ in statements – the *Philosophical Investigations* suggests, and I think correctly, a different evaluation. Recall the key passage in § 242: "If language is to be a means of communication there must be agreement not only in definitions but also (queer as it may sound) in judgements". And in fact: if meaning were nothing but use, if definitions were, thus, essentially *implicit* ones, establishing meanings through stating relations, then idiosyncratic judgments would be bound up not just with idiosyncratic meanings, but, inevitably, with the *loss* of meaning. Writing is the medium in which divergent meanings and judgments can be entertained. But of course this should not lead us to say that written language is a private language. Wittgenstein's definition of the private language problem is so inextricably bound up with his arguments against the possibility of such a language that if one finds his argument implausible one should conclude that the private language problem is, under conditions of literacy, just *uninteresting*, though it might, certainly, regain interest – as literacy peters out.

NOTES

* In presenting this paper I am essentially indebted to the work of G. H. von Wright. As early as 1955, in his "Ludwig Wittgenstein: A Biographical Sketch" *(Philosophical Review*, vol. 64), Professor von Wright suggested that in Wittgenstein's philosophy there are dimensions different from merely those of linguistic analysis; and in his editorial work, Professor von Wright constantly strives to represent the original, chronological, order of composition of Wittgenstein's remarks, thereby enabling the discovery of such logical and psychological patterns in Wittgenstein's thinking that are sometimes veiled by the thematic arrangements Wittgenstein himself or others prepared. I am particularly indebted to Professor von Wright's paper "The Origin and Composition of Wittgenstein's *Investigations*", and to two lists, prepared by his assistant, Mr. André Maury, giving the manuscript sources of the remarks in *Zettel* and *Philosophical Investigations* respectively. A precise formulation of the ideas here presented would have been impossible without the use of these, as yet unpublished, materials.

[1] *On Certainty*, § 132.

[2] In the 1938 Foreword to an early version of the *Philosophical Investigations*. The words are: "in unserem dunkeln Zeitalter", TS 225, p. iii. I am using the catalogue number given in G. H. von Wright, "The Wittgenstein Papers", *The Philosophical Review*, vol. 78 (1969), pp. 483–503.

[3] This is how I interpret the relevant passage in the letter of introduction written by Keynes, upon Wittgenstein's request, to the Soviet Ambassador in London. "I must leave it to [Wittgenstein]", wrote Keynes in that letter, "to tell you his reasons for wanting to go to Russia. He is not a member of the Communist Party, but has strong sympathies with the way of life which he believes the new régime in Russia stands for" (Wittgenstein, *Letters to Russell, Keynes and Moore*, Oxford: Basil Blackwell, 1974, p. 136). The letter in which Wittgenstein formulates his request is dated 6.7.1935. "You would have to say in your introduction", writes Wittgenstein, "that I am your personal friend and that you are sure that I am in no way politically dangerous (that is, if this *is* your opinion). ... I am sure that you partly understand my reasons for wanting to go to Russia and I admit that they are partly bad and even childish reasons but it is true also that behind all that there are deep and even good reasons" (*ibid.*, p. 135).

[4] Paul Engelmann, *Letters from Ludwig Wittgenstein: With a Memoir*, Oxford: Basil Blackwell, 1967, p. 121.

[5] Fania Pascal, "Wittgenstein: A Personal Memoir", *Encounter*, vol. 16, August 1973, p. 25.

[6] Cf. e.g. the following remark: "Austria and the *Austrians* have sunk so miserably low since the war that it's too dismal to talk about". From a letter to Russell, dated

28.11.1921. *Letters to Russell, Keynes and Moore*, p. 98.

[7] Karl Mannheim, "Das konservative Denken", *Archiv für Sozialwissenschaft und Sozialpolitik*, vol. 57 (1927), p. 105. I am indebted to Dr. D. C. Bloor of the University of Edinburgh for directing my attention to the Wittgenstein–Mannheim relation. Cf. his paper "Wittgenstein and Mannheim on the Sociology of Mathematics", *Studies in the History and Philosophy of Science*, vol. 4 (1973). In a letter Dr. Bloor kindly wrote to me in 1974, he suggested that the later Wittgenstein could be regarded as a conservative thinker in the sense defined by Mannheim in the above-quoted essay. I think this is a correct and very fruitful suggestion. I would like to emphasize, however, that my aim in the present paper is not to point out that Wittgenstein was, in the technical sense defined by Mannheim, a conservative, but to show that he was conservative *sans phrase*, and, in particular, that his philosophy provides novel and cogent arguments for a new conservatism or traditionalism.

[8] Wittgenstein mentions Spengler in his comments on Frazer, written in 1931 ("Bemerkungen über Frazers *The Golden Bough*", *Synthese*, vol. 17, 1967, p. 241). In the same year he makes another, highly interesting, reference to Spengler. "Es ist, glaube ich, eine Wahrheit darin", writes Wittgenstein, "wenn ich denke, dass ich eigentlich in meinem Denken nur reproduktiv bin. Ich glaube, ich habe nie eine Gedankenbewegung *erfunden*, sondern sie wurde mir immer von jemand anderem gegeben. Ich habe sie nur sogleich leidenschaftlich zu meinem Klärungsversuch aufgegriffen. So haben mich Boltzmann, Hertz, Russell, Kraus, Spengler, Sraffa beeinflusst" (Wittgenstein, *General Remarks*, compiled by G. H. von Wright, unpublished, vol. II, p. 256). There is, again, an important reference to Spengler in *Philosophische Grammatik*, Part II, written in 1932 or 1933. Discussing some mathematical term, Wittgenstein says that this term, just like other terms, does not have a clear-cut application – it is applied in many, more or less related senses, "[l]ike words such as 'people', 'king', 'religion', etc.; cf. Spengler" (Ludwig Wittgenstein, *Philosophical Grammar*, Oxford: Basil Blackwell, 1974, p. 299).

[9] Some of the stylistic features which first emerge in the *Philosophical Grammar* are: fictitious dialogues, unanswered questions, peculiar and enigmatic similes – and the constant use of "Du".

[10] Cf. A. W. Levi, "Wittgenstein as Dialectician", in: K. T. Fann, ed., *Wittgenstein: The Man and His Philosophy*, New York: Dell, 1967, pp. 372 f.

[11] Cf. R. L. Goodstein, "Wittgenstein's Philosophy of Mathematics", in: Alice Ambrose and Morris Lazerowitz (eds.), *Ludwig Wittgenstein: Philosophy and Language*, London: George Allen and Unwin, 1972.

[12] Here characterized as "a new sort of nomad, cohering unstably in fluid masses, the parasitical city-dweller, traditionless, utterly matter-of-fact, religionless, clever, unfruitful, deeply contemptuous of the peasantry (and especially of its highest form, the landed aristocrat) and thus a fantastic stride towards the unorganic, towards the end" – a type which substitutes a "cold sense of facts for reverence for inherited tradition, for that which is grown" (*The Decline of the West*, New York: 1928, vol. I, p. 32 f.).

[13] "*Rußland*. Die Leidenschaft verspricht etwas. Unser Gerede dagegen ist kraftlos" (Ludwig Wittgenstein, *Schriften*, vol. 3, Frankfurt am Main: Suhrkamp, 1967, p. 142).

[14] *Ibid.*, p. 115.

[15] This is the point that Wittgenstein constantly makes in his comments on Frazer.

[16] *Ludwig Wittgenstein: Philosophy and Language*, pp. 15 f.

[17] *Philosophical Grammar*, p. 96.

[18] *The Blue and Brown Books*, New York: Harper & Row, 1960, p. 3.

[19] Fania Pascal, *op. cit.*, pp. 33 ff.

[20] *Philosophical Investigations*, Part I, § 433. This passage occurs practically word-for-word in the *Philosophical Grammar*, on p. 133.

[21] *Philosophical Investigations*, Part I, § 85.

[22] *Ibid.*, § 186.

[23] *Ibid.*, § 198.

[24] It is in this connection that the term "inexorability" emerges in 1937. Cf. *Remarks on the Foundations of Mathematics*, Oxford: Basil Blackwell, 1964, Part I, §§ 4 and 118.

[25] *Philosophical Investigations*, Part I, § 242.

[26] *On Certainty*, § 156.

[27] *Ibid.*, § 493.

[28] *Ibid.*, §§ 47, 664.

[29] *Ibid.*, § 94.

[30] *Ibid.*, § 449.

[31] *Ibid.*, § 312.

[32] *Ibid.*, § 344.

[33] *Ibid.*, § 559.

[34] Fann, *op. cit.*, p. 35. From the "Autobiography" of Rudolf Carnap.

[35] *Philosophical Investigations*, Part I, § 18.

[36] *Ibid.*, "Preface".

[37] *Remarks on the Foundations of Mathematics*, Part I, Appendix II, § 4.

NOTES TO CHAPTER 2

[1] Cf. Anthony Kenny, *Wittgenstein*, Harmondsworth, Middlesex: Penguin Books, 1973, pp. 139 f.

[2] Compare e.g. the description of the "enduring kernel" of conservatism by Klaus Epstein in his *The Genesis of German Conservatism*, Princeton, N.J.: Princeton University Press, 1966, cf. esp. pp. 13–16, or Gerd-Klaus Kaltenbrunner's analysis of what one could call a *conservative anthropology* (Kaltenbrunner, "Der schwierige Konservatismus", in: Kaltenbrunner, ed., *Rekonstruktion des Konservatismus*, Freiburg i.B.: Rombach, 1972, cf. esp. pp. 45 f.), or indeed Karl Mannheim's "Das konservative Denken", *Archiv für Sozialwissenschaft und Sozialpolitik* 57 (1927).

[3] Cf. Rudolf Haller, "Über das sogenannte Münchhausentrilemma", *Ratio* 16 (1974), pp. 115 and 126 f.

[4] Thus a number of years later Wittgenstein wrote: "Counting (and that means: counting like *this*) is a technique that is employed daily in the most various operations of our lives. And that is why we learn to count as we do: with endless practice, with merciless exactitude; that is why it is inexorably insisted that we shall all say 'two' after 'one', 'three' after 'two', and so on" (*Remarks on the Foundations of Mathematics*, Part I, § 4). This conception of mathematical insight and of the ways in which

arithmetic is learned, is rooted in the same psychological attitude as Wittgenstein's general conception of education. The latter may be illustrated, for example, by his remark: "When you say NO to a child, you should be like a wall and not like a door", cf. K. E. Tranøy, "Wittgenstein in Cambridge 1949–51. Some Personal Recollections", in: *Essays on Wittgenstein in Honour of G. H. von Wright – Acta Philosophica Fennica* 28/1–3, 1976, p. 15.

[5] As Klemens von Klemperer writes: The year 1928 "was the last year of the prosperity which had marked German economy since 1924. ... It was quite clearly an economic and political crisis. ... These were the days when Moeller van den Bruck was read, reread, reedited in popular editions, and all but canonized, when Spengler was eagerly debated... The neo-conservatives were the intellectuals of the Right who pointed toward the long-range spiritual roots of the crisis". (Klemperer, *Germany's New Conservatism: Its History and Dilemma in the Twentieth Century*, Princeton, N.J.: Princeton University Press, 1957, pp. 125 and 118 ff.)

[6] Thomas Mann, "Russische Anthologie", in: Mann, *Rede und Antwort: Gesammelte Abhandlungen und kleine Aufsätze*, Berlin: S. Fischer, 1925, p. 236.

[7] F. M. Dostojewsky, *Die Dämonen*, R. Piper, 1921, pp. XVIII f.

[8] Cf. my essay "Wittgenstein's Later Work in Relation to Conservatism", in: Brian McGuinness, ed., *Wittgenstein and His Times*, Oxford: Basil Blackwell, 1982, pp. 49 ff.

[9] Bolkosky, *The Distorted Image: German Jewish Perceptions of Germans and Germany, 1918–1935*, New York: Elsevier, 1975, p. 49.

[10] Ernst Jünger, "Über Nationalismus und Judenfrage", *Süddeutsche Monatshefte* 27 (Sept. 1930), pp. 843 f.

[11] Wagner, "Das Judentum in der Musik" (1850), in: Wagner, *Gesammelte Schriften und Dichtungen in zehn Bänden* (ed. W. Golther), Berlin: Deutsches Verlagshaus, n.d., vol. V, p. 85.

[12] *Ibid.*, pp. 70 f.

[13] Weininger, *Geschlecht und Charakter: Eine prinzipielle Untersuchung*, 25th ed., Wien: Braumüller, 1923, pp. 431 and 425.

[14] Oswald Spengler, *The Decline of the West*, vol. II, p. 323.

[15] Leipzig: Duncker und Humblot, p. 37.

[16] Weininger, *op. cit.*, p. 422.

[17] Wagner, *loc. cit.*, pp. 79 f.

[18] *Ibid.*, p. 81.

[19] J. P. Hebel, *Werke*, Karlsruhe: 1847, vol. III, p. 207.

[20] *Ibid.*, p. 214.

[21] Quoted in Heidegger, *Aus der Erfahrung des Denkens*, Frankfurt/M.: Klostermann, 1983, p. 142.

[22] Spengler, *loc. cit.*, vol. II, p. 321.

[23] *Ibid.*, p. 322.

[24] Weininger, *op. cit.*, p. 411.

[25] The concern for what is concrete involves relying on examples instead of on a general argument. An entry Wittgenstein set down in his notebook in April 1932 is relevant here: "Ich weiß nicht ob ich es je aufgeschrieben habe, daß ich die Methode, einer grammatischen Betrachtung // einer Betrachtung // eine Anzahl Beispiele

vorzusetzen // voranzustellen // in der Mittelschule von einem Professor namens Heinrich Groag (einem Juden) gelernt habe // ... daß ich die Methode, eine sprachliche Betrachtung mit einer Gruppe von Beispielen zu begründen... //" (MS 113, p. 371).

[26] E. Renan, *Geschichte des Volkes Israel*, transl. E. Schaelsky, Berlin: Cronbach, 1894, vol. I, p. 3.

[27] *Ibid.*, p. 4.

[28] *Ibid.*, p. 6.

[29] Leo Baeck, "Die jüdische Religion in der Gegenwart", *Süddeutsche Monatshefte* 27 (Sept. 1930), pp. 830 f.

NOTES TO CHAPTER 3

[1] "In establishing the system of science", Hempel wrote, "there is a conventional moment... even the singular statements which we adopt, which we regard as true, depend upon which of the formally possible systems we choose. – Our choice is logically arbitrary, but the large number of possibilities for choosing is practically restricted by psychological and sociological factors, as particularly Neurath emphasizes. ... How do we learn to produce 'true' protocol statements? Obviously", writes Hempel referring to Carnap but here in fact going further than Carnap did, "by being conditioned. ... we may say that young scientists are conditioned ... in their university courses... Perhaps the fact of the general and rather congruous conditioning of scientists may explain to a certain degree the fact of a unique system of science" (Carl G. Hempel, "On the Logical Positivists' Theory of Truth", *Analysis* 2, 1935, pp. 52, 56 ff.).

[2] Rudolf Haller, *Fragen zu Wittgenstein und Aufsätze zur Österreichischen Philosophie*, Amsterdam: Rodopi, 1986, pp. 130 ff.

[3] Thus e.g. in Mach's philosophy the role of *tradition*, as regards society in general and science in particular, is, characteristically, depicted in an overwhelmingly negative manner. Handed-down patterns, Mach suggests, "are excellent for soldiers, but they will not fit heads" (Ernst Mach, *Popular Scientific Lectures*, transl. by Thomas J. McCormack, La Salle, Ill.: Open Court, 1986, p. 369). Schools should not "select the persons best fitted for being drilled" (*ibid.*, p. 370) and thereby suppress the "powerful judgment" which "would probably have grown up [in children] if they had learned nothing" (*ibid.*, p. 367). The "aim of instruction" should be "simply to economize on experience" (*ibid.*, p. 191). Even though it is a fact that in science "the majority of the ideas we deal with were conceived by others, often centuries ago" (*ibid.*, p. 196), this does not, according to Mach, represent an essential feature of cognition. It amounts only to an "exquisite economy" (*ibid.*, p. 198): each individual could, in principle, think out everything for himself.

[4] This is in fact confined to three pages in Rudolf Carnap, "Erwiderung auf die vorstehenden Aufsätze von E. Zilsel und K. Duncker", *Erkenntnis* 3 (1932/33). And even here Carnap does not seem to realize the inherent absurdity of the idea of a single individual bringing into existence a scientific theory (cf. *ibid.*, p. 180). Carnap even suggests that it would be an empirical question whether systems brought into existence in this way will dovetail with our own. Science, for Carnap, is not an

essentially collective enterprise; it must be regarded as a happy coincidence that science *can* in fact be pursued in a collective manner.

[5] "We are confined", Neurath writes, "by the conceptions of our surroundings. Moreover the individual has hardly enough power to work himself properly through *one* system, not to speak of several systems", he writes discussing the logically arbitrary nature of any one scientific system ("Radikaler Physikalismus und 'Wirkliche Welt'", 1934, in: Neurath, *Gesammelte philosophische und methodologische Schriften*, ed. R. Haller and H. Rutte, Wien: Hölder-Pichler-Tempsky, 1981, p. 616).

[6] "Anti-Spengler", 1921, cf. Neurath, *Empiricism and Sociology*, ed. M. Neurath and R. S. Cohen, Dordrecht: Reidel, 1973, p. 198.

[7] "Physikalismus", 1931, in: Neurath, *Gesammelte...*, p. 420.

[8] Thomas S. Kuhn, "Reflections on my Critics", in: Imre Lakatos and Alan Musgrave (eds.), *Criticism and the Growth of Knowledge*, Cambridge: Cambridge University Press, 1970, p. 253.

[9] Robert S. Cohen and Thomas Schnelle (eds.), *Cognition and Fact: Materials on Ludwik Fleck*, Dordrecht: Reidel, 1986, p. 66.

[10] Robert K. Merton, "The Sociology of Knowledge", in: Georges Gurvitch and Wilbert E. Moore (eds.), *Twentieth Century Sociology*, New York: The Philosophical Library, 1945, p. 366. The major breakthrough in this area of inquiry, stresses Merton, "consisted of the hypothesis that not only error or illusion or unauthenticated belief but also the discovery of truth was socially (historically) conditioned" (*ibid.*, p. 370).

[11] *Ibid.*, p. 368.

[12] In accordance with Edward L. Schaub, cf. his "A Sociological Theory of Knowledge", *The Philosophical Review* 29 (1920).

[13] Durkheim's "sociologie de la connaissance", writes Hans Joas, "is not a sociology of knowledge which interests itself for macro-sociological connections of dispositions of interest and ideology, but a theory of the social constitution of the fundamental categories of change" (Joas, "Durkheim und der Pragmatismus. Bewußtseinspsychologie und die soziale Konstitution der Kategorien", in: Émile Durkheim, *Schriften zur Soziologie der Erkenntnis*, Frankfurt/M.: Suhrkamp, 1987, p. 267).

[14] Ludwig Gumplowicz, *Grundriß der Sociologie*, Wien: 1885, cf. also Gumplowicz, *Der Rassenkampf: Sociologische Untersuchungen*, Innsbruck: 1883.

[15] Gumplowicz, *Grundriß der Sociologie*, p. 167.

[16] *Ibid.*, pp. 168 f.

[17] *Ibid.*, p. 174. Cf. also *Soziologische Essays*, Innsbruck: 1899, pp. 6, 9, 11.

[18] Émile Durkheim, *Les formes élémentaires de la vie religieuse*, Paris: 1912, p. 303.

[19] "Un homme qui ne penserait pas par concepts", Durkheim writes, "ne serait pas un homme; car ce ne serait pas un être social", *ibid.*, p. 626.

[20] *Ibid.*, p. 14.

[21] *Ibid.*, p. 24.

[22] Cf. Durkheim, *Pragmatisme et sociologie*, Paris: 1981, p. 173.

[23] Maurice Halbwachs, *Les cadres sociaux de la mémoire* (1925), new ed. Paris: 1975, pp. XV ff.

[24] *Ibid.*, pp. 289 f.

[25] *Ibid.*, p. 248.

[26] *Ibid.*, p. 296.

[27] *Ibid.*, p. 291.

[28] Cf. Durkheim, *Les formes...*, pp. 609 and 635.

[29] Ludwik Fleck, *Genesis and Development of a Scientific Fact*, Chicago: University of Chicago Press, 1979, p. 22.

[30] Cf. *ibid.*, p. 10.

[31] *Ibid.*, p. 9. The thesis of Jerzy Giedymin, according to which Fleck's perception of Mach (and of the conventionalists in general) was "largely mistaken" (Cohen and Schnelle, eds., p. 184), is not borne out by the present author's reading of Mach. Nor do "the analogies between the conventionalist philosophy and the sociology of knowledge" Giedymin points to (*ibid.*, pp. 184 f., "sociology of knowledge" here used in the sense of *Wissenssoziologie*) convince in the slightest.

[32] Fleck, *Genesis...*, p. 9. Accordingly, once a closed system of opinions, "consisting of many details and relations", as Fleck puts it, has been formed, it offers "constant resistance" to any criticism (*ibid.*, p. 27).

[33] Cf. *ibid.*, p. 41.

[34] *Ibid.*, p. 39.

[35] Cf. *ibid.*, pp. 42 f. See also the memorable formulation in his essay "Über die wissenschaftliche Beobachtung und die Wahrnehmung im allgemeinen" (1935): "A truly isolated investigator is impossible, and so also is an ahistoric discovery or a styleless observation. An isolated investigator without bias and tradition, without forces of mental society [*Denkgesellschaft*] acting upon him, and without the effect of the education of that society, would be blind and senseless. Thinking is a collective activity, just as choral singing or conversation" (*Cognition and Fact*, p. 77).

[36] Cf. e.g. *Genesis...*, pp. 105 and 107 f.

[37] *Genesis...*, p. 115.

[38] "It is not possible that there should have been only one occasion on which someone obeyed a rule. It is not possible that there should have been only one occasion on which a report was made, an order given or understood; and so on. – To obey a rule, to make a report, to give an order, to play a game of chess, are *customs* (uses, institutions). – To understand a sentence means to understand a language. To understand a language means to be master of a technique" (Wittgenstein, *Philosophical Investigations*, Part I, § 198).

[39] Cf. esp. G. P. Baker and P. M. S. Hacker, *Wittgenstein: Rules, Grammar and Necessity*, Oxford: Basil Blackwell, 1985, p. 242.

[40] Wittgenstein, *On Certainty*, Oxford: Basil Blackwell, 1969, § 140.

[41] Durkheim, *Les formes...*, p. 23.

[42] Halbwachs, *op. cit.*, p. 20.

[43] *On Certainty*, §§ 343 and 558.

[44] Durkheim, *Les formes...*, p. 19.

[45] Halbwachs, *op. cit.*, p. 145.

[46] Wittgenstein, *Bemerkungen über die Grundlagen der Mathematik*, Frankfurt/M.: Suhrkamp, 1974, p. 195.

[47] Durkheim, *Les formes...*, p. 25.

[48] *Bemerkungen über die Grundlagen der Mathematik*, pp. 80 f. This is a view

Wittgenstein obviously developed under the influence of the Marxist Piero Sraffa. Thus consider MS 113, pp. 318 f., written towards the end of 1931 (cf. also TS 211, pp. 572 f.): "(Sraffa) Ein Ingenieur baut eine Brücke; er schlägt dazu in mehreren Handbüchern nach; in technischen Handbüchern und in juridischen. Aus den einen erfährt er, daß die Brücke zusammenbrechen würde, wenn er diesen Teil schwächer machen würde als etc. etc.; aus den andern daß er eingesperrt würde, wenn er sie so und so bauen wollte // würde //. – Stehn nun die beiden Bücher nicht auf gleicher Stufe? – Das kommt drauf an, was für eine Rolle sie in seinem Leben spielen". ("An engineer builds a bridge; he checks in various handbooks; in technical handbooks and in legal ones. He learns from the one that the bridge would collapse if he were to make this part weaker than etc. etc.; from the other he learns that he would be jailed if he wanted to build it in such-and-such a way. – Do these two books now stand on the same level? – That depends upon what sort of role they play in his life".)

49 Halbwachs, *op. cit.*, p. 62.
50 *Cognition and Fact*, p. 79.
51 *On Certainty*, §§ 47, 161, 283, 298.
52 Halbwachs, *op. cit.*, p. 59.
53 Fleck, *Genesis*, p. 48.
54 Wittgenstein, *Philosophical Investigations*, § 5.
55 The "divers modes d'association des souvenirs résultent des divers façons dont les hommes peuvent s'associer. On ne comprend bien chacun d'eux, tel qu'il se présente dans la pensée individuelle, que si on le replace dans la pensée du groupe correspondant. On ne comprend bien quelle est leur force relative, et comment il se combinent dans la pensée individuelle, qu'en rattachant l'individu aux groupes divers dont il fait en même temps partie" (Halbwachs, *op. cit.*, p. 144). The "creative expert [is] the personified intersection of various thought collectives as well as of various lines of development of ideas" (Fleck, *Genesis*, p. 118). This view of creativity is developed at length in Barry Smith, "Practices of Art", in: J. C. Nyíri and B. Smith (eds.), *Practical Knowledge. Outlines of a Theory of Traditions and Skills*, London: Croom Helm, 1988, cf. esp. pp. 192–99.
56 *Zettel*, § 323.
57 *Op. cit.*, p. XVI.
58 Cf. chapter 2, p. 11 above, see also Wittgenstein, *Philosophical Investigations*, §§ 95, 110, 427 f.
59 Halbwachs, *op. cit.*, p. 53; Wittgenstein, *Philosophical Investigations*, § 329.
60 Halbwachs, *op. cit.*, p. 57.
61 Wittgenstein, *Philosophical Investigations*, § 202, cf. also §§ 258 ff.
62 Halbwachs, *op. cit.*, pp. 21 f.
63 *Ibid.*, p. 39.
64 Halbwachs, *op. cit.*, pp. 62 f. Cf. also Wittgenstein, *Philosophical Investigations*, § 139.
65 *Op. cit.*, p. 69.
66 Cf. *Zettel*, § 371.
67 Ludwig Wittgenstein, *The Blue and Brown Books*, Oxford: Basil Blackwell, 1958, p. 93.
68 *Ibid.*, p. 94.

[69] *Cognition and Fact*, p. 145.
[70] *Philosophical Investigations*, § 88.
[71] *Cognition and Fact*, p. 136.
[72] *Philosophical Investigations*, Part II, sect. xi.
[73] *Cognition and Fact*, p. 108.
[74] *Bemerkungen über die Grundlagen der Mathematik*, p. 99.
[75] "The importance of sociological methods in the investigation of intellectual activities", Fleck writes, "was already recognized by Auguste Comte. Recently it was stressed by Durkheim's school in France and by the philosopher Wilhelm Jerusalem among others in Vienna. – Durkheim speaks expressly of the force exerted on the individual by social structures... He also mentions the superindividual and objective character of ideas belonging to the collective [*Kollektivvorstellungen*]" (*Genesis*, p. 46).
[76] Durkheim reviewed the book of Gumplowicz, *Grundriß der Sociologie*, in 1885; Gumplowicz mentions Durkheim's *Règles* in his essay "Das Eigenthum als soziale Thatsache" in 1895 (cf. Gumplowicz, *Soziologische Essays*, p. 67).
[77] Cf. Fleck, *Genesis*, pp. 46 f.
[78] *Genesis...*, p. 37, n. 21, and p. 47.
[79] *Genesis...*, pp. 47 ff., cf. also *Cognition and Fact*, p. 80. – Fleck also refers to Simmel's *Soziologie*, to Le Bon's *Psychologie des foules*, to McDougall's *The Group Mind* and to Freud's "Massenpsychologie und Ich-Analyse", pointing out that Le Bon "deals almost exclusively with the momentary mass, mainly in its violent state of emotion" and thus "sees in any socialization merely a degradation of psychological qualities", while both McDougall and Freud are unable to "explain the specific, nonadditive elements of the mass psyche" (*Genesis...*, pp. 179 f., n. 7). The notion of "style", in the sense of a "specific type of thinking", which occurs already in the very first sentence of Fleck's first philosophical paper (published in 1927 – cf. *Cognition and Fact*, p. 39), is obviously derived from Spengler. The hypothesis that Polish philosophy in the inter-war period might have had some influence on Fleck is plausibly argued for by Jerzy Giedymin (in *Cognition and Fact*, pp. 179–215) but is vastly exaggerated by Thomas Schnelle (cf. esp. Schnelle, *Ludwik Fleck – Leben und Denken: Zur Entstehung und Entwicklung des soziologischen Denkstils in der Wissenschaftsphilosophie*, Freiburg: Hochschulverlag, 1982, and Schnelle in *Cognition and Fact*, pp. 14–19). As Schnelle puts it, "we may assume it to be certain" that Fleck was in contact with "the three outstanding philosophers in Lwów", *viz.* Twardowski, Ajdukiewicz and Chwistek, *even though* "Fleck never explicitly mentions them" (*Cognition and Fact*, pp. 18 f.), and even though, as Schnelle of course himself realizes, the position developed by Fleck ran counter to their respective positions. In fact the only direct proof of Fleck's "Polish connection" is an offprint of an essay of his, bearing the dedication: "To the right honorable Professor Ajdukiewicz, given by the author, who requests his kind acceptance. – 20.7.35". (*Cognition and Fact*, p. 16) And not even Schnelle maintains that Polish philosophy could have been in any way instrumental in building up Fleck's *sociological perspective* as regards epistemological issues – the very point which needs to be pursued.
[80] Cf. Katerina Clark and Michael Holquist, *Mikhail Bakhtin*, Cambridge, Mass.:

Harvard University Press, 1984, pp. 19 f. Durkheim's spiritual presence in Paris in the 1920s and 1930s was still "overwhelming", as René König writes, recalling his own student years there (König, "Émile Durkheim", in: Dirk Käsler, ed., *Klassiker des soziologischen Denkens*, vol. 1, München: C. H. Beck, 1976, p. 353).

[81] Steven Lukes, *Émile Durkheim: His Life and Work*, London: Allen Lane, 1973, p. 39.

[82] *Ibid.*, p. 40.

[83] Gumplowicz, *Der Rassenkampf*, p. 334.

[84] William M. Johnston, *The Austrian Mind: An Intellectual and Social History, 1848–1938*, Berkeley: University of California Press, 1972, p. 323.

[85] Otto Neurath, *Empiricism and Sociology*, p. 41.

[86] *Ibid.*, pp. 2 ff.

[87] *Ibid.*, p. 81.

[88] The "stranger" being one of his technical concepts, with which he has been held to have made "a significant contribution to the sociology of the Jews" (Werner J. Cahnman, "Sociology", in: *Encyclopaedia Judaica*, Jerusalem: n.d., p. 65).

[89] In 1908 Simmel, who at that time taught in Berlin, was considered for a chair of philosophy at Heidelberg. At the request of the ministry concerned, Dietrich Schaefer wrote an evaluation of the candidate in which he says: "I will express my opinion about Professor Simmel quite frankly. I do not know whether or not he has been baptized... He is, at any rate, a dyed-in-the-wool Israelite, in his outward appearance, in his bearing, and in his manner of thinking... a Semitic lecturer – wholly, partially, or philo-Semitic, whatever he is – will find fertile soil, no matter what the circumstances, at a university where the corresponding part of the student body numbers several thousand, given the cohesion that prevails in these circles. ... it is impossible for me to believe that the level of Heidelberg would be raised by allowing even broader space than it already occupies among the faculty to the world view and philosophy of life which Simmel represents, and which, after all, are only too obviously different from our German Christian-classical education. ... Simmel owes his reputation chiefly to his 'sociological' activity. ... It is my view, however, that sociology has yet to earn its position as a scholarly discipline. It is, in my opinion, a most perilous error to put 'society' in the place of state and church as the decisive organ of human coexistence", (Lewis A. Coser, "Georg Simmel's Style of Work: A Contribution to the Sociology of the Sociologist", *The American Journal of Sociology* 63 [1957–1958], pp. 640 f.)

[90] Cf. Georg Lukács, *Dostojewski: Notizen und Entwürfe*, ed. J. C. Nyíri, Budapest: Akadémiai, 1985, see esp. pp. 156 ff., 182 f., 186 f., and 189 f.

[91] For an excellent description of the intellectual preeminence of Jews in modern Austria see Johnston, *op. cit.*, pp. 23 ff. For an analysis of the intellectual dynamics resulting from Jewish emancipation compare also William O. McCagg, Jr., *Jewish Nobles and Geniuses in Modern Hungary*, Boulder: East European Quarterly, 1972.

[92] Émile Durkheim, *Le Suicide: Étude de Sociologie*, Paris: 1897, pp. 159 f.

[93] Leo Baeck, *Das Wesen des Judentums*, 6th ed., Köln: Melzer, 1960, pp. 299 f.

[94] "The duty to research has brough it about in Jewry that the inherited contents of teaching are not taken over as finished and complete, but rather constantly renewed in the consciousness of the community", *ibid.*, p. 19.

[95] *Ibid.*, pp. 51 f.

[96] Martin Buber, *Der Jude und sein Judentum: Gesammelte Aufsätze und Reden*, Köln: Melzer, 1963, pp. 188 ff.

[97] Baeck, *op. cit.*, pp. 210 ff.

[98] "... whoever has accepted the religion of Israel can and should be seen as the son of Abraham" (*ibid.*, p. 80), i.e. common faith constitutes, rather than depends on, common ancestry – a clear case indeed of the past being constructed rather than simply given.

[99] For him, accordingly, Jewish consciousness is but "racial pride", and he agrees with the view held by his father that "living in an overwhelmingly Christian society impose[s] the obligation to give as little offence as possible – to become assimilated" (Karl R. Popper, *Unended Quest: An Intellectual Autobiography*, Fontana, 1976, p. 105).

[100] Compare the contrast Lukes refers to with respect to Durkheim and Max Weber: "For Durkheim, sociological explanation involved seeking 'immediate and determining [social] causes'; for Weber, 'subjective understanding is the specific characteristic of sociological knowledge'... For Durkheim, social phenomena are *sui generis* realities... for Weber, 'these collectives must be treated as *solely* the resultants and modes of organization of the particular acts of individual persons, since these alone can be treated as agents in a course of subjectively understandable action'" (Lukes, *op. cit.*, pp. 19 f.).

[101] The connection with the *Volksgeist* of the German Romantics is, then, a merely apparent one. No wonder that, while Fleck's book did not make an impact on the logical positivists at the time it was first published, it was not taken up, either, by philosophical circles increasingly dominated by Germanic thought. I cannot agree with Wolfgang Stock's conclusion that "it will be possible with a few supplementary assumptions to modify the Fleckian theory so that it is consistently integratable into the theories of the national-socialist/racist ideology" (Wolfgang S. Stock, "Die Bedeutung Ludwik Flecks für die Theorie der Wissenschaftsgeschichte", *Grazer Philosophische Studien* 10, 1980, p. 113; that Fleck cannot be charged with invoking any mysterious, superorganic group mind was clearly shown by Douglas in her synoptic treatment of Durkheim and Fleck. Cf. Mary Douglas, *How Institutions Think*, Syracuse, N.Y.: Syracuse University Press, 1986, p. 14). And that Fleck's book was then rediscovered, or rather discovered, after the war, is not just a matter of a changed intellectual climate, as W. Baldamus suggests (cf. W. Baldamus, "Ludwig Fleck and the Development of the Sociology of Science", in: Peter R. Gleichmann, Johan Goudsblom, and Hermann Korte, eds., *Human Figurations: Essays for Norbert Elias*, Amsterdam: Stichting Amsterdams Sociologisch Tijdschrift, 1977, p. 151), but also a matter of change in the sphere of *politics* and of *ideology*.

[102] Thus, most notably, in Johnston, *op. cit.*

[103] Cf. Barry Smith, "Austrian Origins of Logical Positivism", in: B. Gower (ed.), *Logical Positivism in Perspective*, London: Croom Helm, 1987, pp. 46 ff.

NOTES TO CHAPTER 4

[1] Cf. his letter of Aug. 5, 1890, to Conrad Schmidt – "wenn die materielle Daseinsweise das primum agens ist, [schließt] das nicht aus, daß die ideellen Gebiete eine reagierende, aber sekundäre Einwirkung auf sie hinwiederum ausüben" – or, e.g., his letter of Sept. 21/22 of the same year to Joseph Bloch: "Wir machen unsere Geschichte selbst, aber ... unter sehr bestimmten Voraussetzungen und Bedingungen. Darunter sind die ökonomischen die schließlich entscheidenden. Aber auch die politischen usw., ja selbst die in den Köpfen der Menschen spukende Tradition, spielen eine Rolle..."

[2] The relevant classic essay being, of course, his "Die protestantische Ethik und der 'Geist' des Kapitalismus", *Archiv für Sozialwissenschaft und Sozialpolitik* XX–XXI (1903–1904).

[3] I am referring to his fateful early notion of class consciousness, cf. esp. the essay "Klassenbewußtsein" in his *Geschichte und Klassenbewußtsein* (1923).

[4] Moscow: n.d., p. 122.

[5] "Die Produktionsweise des materiellen Lebens bedingt den sozialen, politischen und geistigen Lebensprozess überhaupt". *Zur Kritik der politischen Ökonomie*, Marx–Engels, *Werke*, vol. 13, Berlin: Dietz, 1971, pp. 8 f.

[6] Recall the view succinctly expressed by Hayek in his 1945 talk "Individualism: True and False", where he spoke about "true individualism" affirming the value of the family, the local community, and of "common conventions and traditions" (Hayek, *Individualism and Economic Order*, 1949, London: Routledge & Kegan Paul, 1976, p. 23). Hayek called into question "whether a free or individualistic society can be worked successfully if people are too 'individualistic' in the false sense, if they are too unwilling voluntarily to conform to traditions and conventions" (*ibid.*, p. 26). Or, as Eisenstadt has put it: "the implicit assumption in many studies (and the one most closely related to the dichotomous conception of traditional versus modern societies) that the less 'traditional' a society is, the more capable it is of sustained growth was proven incorrect. It became clear that the mere destruction of traditional forms did not necessarily assure the development of a new, viable, modern society... In addition ... it was realized that in some countries, such as Japan or England, modernization had been successfully undertaken under the aegis of traditional symbols and even traditional elites... ... a growing array of evidence has shown that many forms of extended-family and/or kinship relations may indeed not only be compatible with industrialization but even reinforce it" (S. N. Eisenstadt, *Tradition, Change, and Modernity*, New York: Wiley, 1973, pp. 98 f. and 108).

[7] "Wir kennen jetzt die *Substanz* des Werts. Es ist die *Arbeit*. Wir kennen sein *Größenmaß*. Es ist die *Arbeitszeit*". *Das Kapital*, vol. 1, Berlin: Dietz, 1969, p. 55.

[8] "Als Tauschwert sind alle Waren nur bestimmte Maße *festgeronnener Arbeitszeit*", *Zur Kritik der Politischen Ökonomie*, p. 18.

[9] "Knowledge" and "information" are of course not identical; but they are closely related. As Fred I. Dretske puts it in his *Knowledge and the Flow of Information*, Oxford: Basil Blackwell, 1981, p. 44: "Roughly speaking, information is that commodity capable of yielding knowledge, and what information a signal carries is what we can learn from it". – "By information", writes Daniel Bell, "I mean data

processing in the broadest sense; the storage, retrieval, and processing of data becomes the essential resource for all economic and social exchanges. ... By knowledge, I mean an organized set of statements of facts or ideas, presenting a reasoned judgment or an experimental result, which is transmitted to others through some communication medium in some systematic form" ("The Social Framework of the Information Society", in M. L. Dertouzos and Joel Moses, eds., *The Computer Age: A Twenty-Year View*, Cambridge, Mass.: MIT Press, 1979, p. 168).

[10] "[W]ith the shortening of labor time and the diminution of the production worker ... it becomes clear that knowledge and its applications replace labor as the source of 'added value' in the national product. In that sense, just as capital and labor have been the central variables of industrial society, so information and knowledge are the crucial variables of postindustrial society", Bell, *loc. cit.*, p. 168.

[11] In his 1857–58 manuscripts, later published as *Grundrisse der Kritik der Politischen Ökonomie*, he wrote: "to the degree that large industry develops, the creation of real wealth comes to depend less on labour time and on the amount of labour employed than on the power of the agencies set in motion during labour time, whose 'powerful effectiveness' is itself in turn out of all proportions to the direct labour time spent on their production, but depends rather on the general state of science and on the progress of technology... As soon as labour in the direct form has ceased to be the great well-spring of wealth, labour time ceases and must cease to be its measure", *Foundations of the Critique of Political Economy*, transl. by Martin Nicolaus, Penguin Books, 1973, pp. 704 f.

[12] Cf. David McLellan, *Karl Marx: His Life and Thought*, London: Macmillan, 1973, p. 299. In another sense this theory is irrefutable since it is from the start *circular*. Consider Marx's central introductory argument: "Um die Tauschwerte der Waren an der in ihnen enthaltenen Arbeitszeit zu messen, müssen die verschiedenen Arbeiten selbst reduziert sein auf unterschiedslose, gleichförmige, einfache Arbeit, kurz auf Arbeit, die qualitativ dieselbe ist und sich daher nur quantitativ unterscheidet. – Diese Reduktion erscheint als eine Abstraktion, aber es ist eine Abstraktion, die in dem gesellschaftlichen Produktionsprozeß täglich vollzogen wird", *Zur Kritik der Politischen Ökonomie*, p. 18.

[13] Cf. *Foundations*, p. 162. I am here quoting after the English translation as cited by McLellan, *op. cit.*, p. 297.

[14] In a telling passage in one of his earliest writings he regards it as a "correct idea" that early, primitive conditions in a way foreshadow the genuine conditions under which man ought to live, "daß die *rohen* Zustände naive niederländische Gemälde der *wahren* Zustände sind" ("Das philosophische Manifest der historischen Rechtsschule", 1842, Marx–Engels, *Werke*, vol. 1, Berlin: Dietz, 1964, p. 78).

[15] Cf. *Foundations*, pp. 325 and 611: "the development of the productive powers of labour" makes it possible that "the possession and preservation of general wealth require a lesser labour time of society as a whole, and ... the labouring society relates scientifically to the process of its progressive reproduction, its reproduction in a constantly greater abundance; hence where labour in which a human being does what a thing could do has ceased". Here, then, emerge "the material elements for the development of the rich individuality which is as all-sided in its production as in its consumption, and whose labour ... therefore appears no longer as labour, but as the

full development of activity itself, in which natural necessity in its direct form has disappeared...". It is at this stage that labour truly becomes *travail attractif*, "self-realization, objectification of the subject, hence real freedom, whose action is, precisely, labour...".

[16] *The Second Self: Computers and the Human Spirit*, London: Granada, 1984, p. 173. *"Structured programming"*, the programming mode or style which is sufficiently constrained to permit smooth team co-operation, is "good for business, death for the joy of work", as one programmer interviewed by Turkle expressed the matter (*ibid.*). For a more detailed discussion of structured programming, in the spirit of Adam Smith, cf. David Levy, "Constraining the Choice Set: Lessons from the Software Revolution", *Reason Papers*, Spring 1985, pp. 77–88.

[17] *Turing's Man: Western Culture in the Computer Age*, Chapel Hill: The University of North Carolina Press, 1984, pp. 231 f.

[18] *Ibid.*, pp. 232 f.

[19] The text she directly bases her analyses on seems to be ch.13 of *Das Kapital*: "Maschinerie und große Industrie". But, here again, the *Foundations* is worth quoting too: "The science which compels the inanimate limbs of the machinery, by their construction, to act purposefully, as an automaton, does not exist in the worker's consciousness, but rather acts upon him through the machine as an alien power, as the power of the machine itself" (p. 693).

[20] *Op. cit.*, p. 172.

[21] *Ibid.*, pp. 172 f.

[22] *Ibid.*, pp. 186 f.

[23] Freedom within the sphere of material production, Marx writes in volume three of *Das Kapital*, "kann nur darin bestehn, daß der vergesellschaftete Mensch, die assoziierten Produzenten ... ihren Stoffwechsel mit der Natur rationell regeln, unter ihre gemeinschaftliche Kontrolle bringen... ... Aber es bleibt dies", *viz.* the sphere of production, "immer ein Reich der Notwendigkeit. Jenseits derselben beginnt die menschliche Kraftentwicklung, die sich als Selbstzweck gilt, das wahre Reich der Freiheit, das aber nur auf jenem Reich der Notwendigkeit als seiner Basis aufblühen kann. Die Verkürzung des Arbeitstags ist die Grundbedingung", Berlin: Dietz, 1969, p. 828.

[24] Cf. Klaus Haefner, *Die neue Bildungskrise: Lernen im Computerzeitalter*, Reinbek bei Hamburg: Rowohlt, 1985, pp. 197 and 245 f. Haefner registers a certain similarity between what he calls the "humanely computerized society" of the future on the one hand, and ancient Greece on the other. "Bei den Griechen", he writes, "gelang es, Demokratie, Künste und Kultur auf der Basis intensiver Nutzung der Sklaven zu entwickeln; manuelle und kognitive Routinearbeit konnten vergeben werden, dem freien Griechen blieb das 'reine Denken' (und das Kriegshandwerk). Die human computerisierte Gesellschaft nutzt die 'intelligente' Informationstechnik in der Kombination mit dem hohen Stand von Agrar-, Konsumgüter- und Energietechnik zur Lösung vieler 'Sklavenprobleme' im täglichen Leben...", *ibid.*, p. 240. Comparing computers to slaves is, incidentally, not lacking a certain epistemological and ethical interest. I have touched on this issue in my paper "Wittgenstein and the Problem of Machine Consciousness", *Grazer Philosophische Studien* 33/34 (1989), pp. 375–394.

[25] Alvin Toffler, *The Third Wave*, New York: Bantam Books, 1980, p. 194.

[26] *Ibid.*, p. 205. Indeed Toffler speaks of "a new civilization" which "begins to heal the historical breach between producer and consumer, giving rise to the 'prosumer' economics of tomorrow. For this reason, among many, it could ... turn out to be the first truly humane civilization in recorded history", *ibid.*, p. 11. Another aspect of "prosumer" economy is what Michael L. Dertouzos calls "individualized automation", the technological possibility of producing, at home, with the help of intelligent tools, commodities for one's own use, cf. Dertouzos – Moses, eds., *The Computer Age*, pp. 38 ff.

[27] Toffler, *op. cit.*, p. 204.

[28] Turkle, *op. cit.*, pp. 174 f.

[29] Applied to the field of politics this is tantamount to saying that participatory democracy can be deepened by making use of computer networks – not so much in order to count votes, as rather to disseminate relevant knowledge. "The genius of democratic government is not arithmetic", writes Herbert A. Simon, "it is informed consensus. ... The computer enters as a tool that permits policy alternatives to be examined with a sophistication and explicitness that would otherwise be impossible" (Simon, "The Consequences of Computers for Centralization and Decentralization", in Dertouzos – Moses, eds., *The Computer Age*, pp. 224 f.).

[30] Cf. the pioneering early essay by the Hungarian historian István Hajnal, "Le rôle social de l'écriture et l'évolution européenne", *Revue de l'Institut de Sociologie* (Bruxelles), 1934.

[31] Man is today, as Haefner puts it, "hineingestellt in eine mit seinem Gehirn allein *nicht mehr überschaubare informationelle Umwelt...* Die Informationsexplosion ... hat den einzelnen Menschen in eine relativ willkürliche Ecke seiner informationellen Umwelt geschleudert...", *op. cit.*, pp. 31 f.

NOTES TO CHAPTER 5

[1] Gilbert Ryle, *The Concept of Mind*, London: Hutchinson's University Library, 1949, p. 59.

[2] Michael Polanyi, *Science, Faith and Society* (1946). Enl. ed., Chicago: University of Chicago Press, 1964, p. 14. The quote is from the 1964 Introduction, but similar passages occur in the main body of the text, e.g. on pp. 42 f. and 76.

[3] Michael Polanyi, *Personal Knowledge: Towards a Post-Critical Philosophy* (1958). London: Routledge & Kegan Paul, 1983, p. 49.

[4] *Ibid.*, p. 53. Similar formulations can be found in John Ziman's *Public Knowledge: An Essay concerning the Social Dimension of Science*, Cambridge: Cambridge University Press, 1968, e.g. on pp. 7 and 10: "The fact is that scientific investigation ... is a practical art. It is not learnt out of books, but by imitation and experience. ... The young scientist does not study formal logic, but he learns by imitation and experience a number of conventions that embody strong social relationships".

[5] Michael Oakeshott, *Rationalism in Politics*. London: Methuen, 1962, pp. 102 f. The quote is from the essay "Rational Conduct" (1950).

[6] "Reflections on my Critics", p.275, in: Imre Lakatos and Alan Musgrave, eds., *Criticism and the Growth of Knowledge*, Cambridge: Cambridge University Press, 1970.

[7] David Bloor, "The Strengths of the Strong Programme" (1981), repr. in: J. R. Brown, ed., *Scientific Rationality: The Sociological Turn*, Dordrecht: Reidel, 1984, p. 88. The parallels (and differences) between Oakeshott and Kuhn are illuminatingly brought out in M. D. King, "Reason, Tradition, and the Progressiveness of Science" (1971), repr. in Gary Gutting, ed., *Paradigms and Revolutions: Appraisals and Applications of Thomas Kuhn's Philosophy of Science*, Notre Dame: University of Notre Dame Press, 1980. Kuhn, King writes, "states emphatically that the term 'paradigm' denotes not a world-view but a specific example of actual scientific practice which serves as a model for a research community and implicitly defines the legitimate problems and methods of a research field for successive generations of practitioners. ... Faithfulness to the traditions which spring from paradigms or sets of paradigms is the hallmark of genuine 'science'. To break faith with established tradition is to risk being labelled a crank, a charlatan, or being made an 'outlaw'. – A sociologist reading Kuhn's attack on scientific rationalism can hardly fail to be struck by how closely it resembles Oakeshott's famous onslaught against political rationalism; Kuhn's science like Oakeshott's politics is subject to authority of concrete traditions rather than that of abstract 'reason'. Both are seen as practical activities that, to use Oakeshott's distinction, involve not merely technical knowledge (or technique) which 'is susceptible of formulation in rules, principles, directions, and maxims' and which may therefore be learned from a book and thereafter 'applied', but also practical knowledge which cannot be reduced to rules, cannot be written down and therefore 'can neither be taught nor learned, but only imparted'" (Gutting, ed., pp. 108 f.; the Oakeshott reference is to his "Rationalism in Politics", 1947).

[8] Paul Feyerabend, *Science in a Free Society* (1978). London: Verso, 1982, p. 26.

[9] G. H. von Wright, *Wittgenstein*, Oxford: Basil Blackwell, 1982, p. 178.

[10] Cf. Rudolf Haller, *Urteile und Ereignisse: Studien zur philosophischen Logik und Erkenntnistheorie*, Freiburg: Verlag Karl Alber, 1982, p. 184.

[11] The crucial passages being *Philosophical Investigations* §§ 85, 198–208, 239–242.

[12] As *one* strand in the thought of Plato already seems to suggest: see the reconstruction in Wolfgang Wieland, *Platon und die Formen des Wissens*, Göttingen: Vandenhoeck & Ruprecht, 1982, esp. p. 254: "Of course knowledge of this kind" – e.g. the expert knowledge possessed by craftsmen – "will be transmitted always only through a process of instruction and practice. It will never be capable of being transferred like an object. It is paradigmatic of the knowledge of the craftsman that he who possesses it cannot distance himself from it... It cannot be objectified, because – as a happy metaphor of Plato's has it – it is as it were grown into the action itself". It is characteristic of the inner tensions in Plato's philosophy that at the same time he is of course the very originator of the idea of *abstract* knowledge, cf. chapter 9, p. 96.

[13] *Personal Knowledge*, pp. 49 f.

[14] *Ibid.*, pp. 54 f.

[15] Ludwik Fleck, *Genesis and Development of a Scientific Fact*, Chicago: The University of Chicago Press, 1979, p. 53. Cf. Ludwik Fleck, *Entstehung und Entwicklung einer wissenschaftlichen Tatsache: Einführung in die Lehre vom Denkstil und Denkkollektiv* (1935), Frankfurt/M.: Suhrkamp, 1980, p. 72. Incidentally, the notion of practical knowledge is, in modern literature, foreshadowed in the work of Max Scheler, who presumably had some, direct or indirect, influence on Fleck (cf.

Fleck, *Entstehung*, p. 64, n. 29). As Scheler wrote in his "Der Formalismus in der Ethik und die materiale Wertethik" (1913): "There is something like 'practical' obeying and 'disobeying' of laws, but not of laws which 'control' natural acting as natural laws control, in the sense that natural acting would conform 'to' them in an objective manner. The laws that we have in mind are not at all given as laws (in a form of perception, of 'being conscious of...'); they are *experienced* as fulfilled or broken in the execution of acting. And it is only in these experiences that they are given. In this sense the acting artist is 'controlled' by the aesthetic laws of his art without 'applying' them; nor does he realize their fulfillment or violation only in the effect, i.e., in the work of art produced. In this sense, too, it belongs to the essence of the 'crime' that he who breaks laws experiences himself as breaking them while acting; these are laws with which he reckons in *practice*, whether he or others are concerned, without having to have the slightest *knowledge* of such laws, and without having to have 'thought' about them". (Quoted from Scheler, *Formalism in Ethics and Non-Formal Ethics of Values*, transl. M. S. Frings and R. L. Funk, Evanston: Northwestern University Press, 1973, pp. 141 f. The notion approximated here is of course *not* the *"practical-technical intelligence"* described e.g. in Scheler, *Problems of a Sociology of Knowledge*, London: Routledge & Kegan Paul, 1980, cf. esp. p. 79.)

[16] *Genesis...*, p. 53.

[17] Edward A. Feigenbaum and Pamela McCorduck, *The Fifth Generation*, New York: Signet, 1984, p. 67.

[18] *Ibid.*, p. 82.

[19] *Critique of Pure Reason*, A 132–134.

[20] *The Concept of Mind*, pp. 29 f.

[21] Cf., e.g., *Philosophical Investigations*, §§82–86 and 198 ff.

[22] *Science, Faith and Society*, p. 14.

[23] Hayek, *Studies in Philosophy, Politics and Economics*, London: Routledge & Kegan Paul, 1967, p. 62.

[24] *The Concept of Mind*, p. 56.

[25] *Ibid.*, p. 26. Similarly Feyerabend: "What is called 'reason' and 'practice' are ... two different types of practice", *Science in a Free Society*, p. 26. Also Arnold Gehlen, even if on the basis of some rather crude arguments: "Das menschliche Erkennen ... ist geradezu als Phase der Handlung zu definieren", *Der Mensch*, Berlin: Junker und Dünnhaupt, 1940, p. 52.

[26] See e.g. the discussion in Samuel Coleman, "Is There Reason in Tradition?", in: Preston King and B. C. Parekh, eds., *Politics and Experience*, Cambridge: Cambridge University Press, 1968, cf. esp. pp. 242 ff.

[27] Karl R. Popper, *Conjectures and Refutations*, London: Routledge and Kegan Paul, 1972 printing, pp. 130 f. From the paper "Towards a Rational Theory of Tradition" (1948).

[28] *Rationalism in Politics*, p. 109. Recently the same point was made by O. Schwemmer. One participates, writes Schwemmer, in the *"Handlungskultur"*, i.e. in the universally available forms of activity, of a given group or society; and by the very possibility of such participation the rational character of those forms is established: "the capacity of he who acts of being able to act in a way intelligible to others ... thereby lends his actions an elementary rationality" (Oswald Schwemmer, "Aspekte

der Handlungsrationalität. Überlegungen zur historischen und dialogischen Struktur unseres Handelns", in: H. Schnädelbach, ed., *Rationalität*, Frankfurt/M.: Suhrkamp, 1984, p. 191).

[29] *Science in a Free Society*, p. 7.

[30] *Rationalism in Politics*, p. 102.

[31] Paris: 1925. In Peter L. Berger's and Thomas Luckmann's *The Social Construction of Reality*, a book which amply stresses the significance of the "pretheoretical level" of knowledge in society (e.g. on p. 65 of the 1967 Anchor Books edition), mention is made of Halbwachs' category of "collective memory" (*ibid.*, p. 202) – but not of his combining "memory" and "tradition" with reason.

[32] Quoted from the German edition: *Das Gedächtnis*, Berlin: Luchterhand, 1966, pp. 348 f. and 383. Halbwachs' suggestion actually represents a third way between the usual alternatives of either equating rationality with an attitude having some unique, standard structural characteristics, an attitude marred only by false logic, traditions, and emotions; or by accepting as rational any views or positions that are felt, by the groups or persons holding them, to be appropriate under the obtaining circumstances. These are the two alternatives called – rather misleadingly – the *"traditionelle Rationalitätskonzeption"* and the *"anti-traditionalistisches Rationalitätskonzept"* by Karl Acham, in his essay "Über einige Rationalitätskonzeptionen in den Sozialwissenschaften", in: H. Schnädelbach, ed., *Rationalität*.

[33] The essay was first published in 1917. Quoted from Eliot, *The Sacred Wood: Essays on Poetry and Criticism* (1920), London: Methuen, 1960, pp. 56–59.

[34] Edmund Burke, "A Philosophical Inquiry into the Origin of our Ideas of the Sublime and Beautiful" (1756/57), in: Burke, *The Works: Twelve Volumes in Six*, 1887, vol. I, pp. 246–252.

[35] Halbwachs, *Das Gedächtnis*, p. 355.

[36] Polanyi, *Personal Knowledge*, pp. 207 f.

[37] Cf. chapter 3 above, p. 32.

[38] Wittgenstein, *Culture and Value*, p. 76

[39] Burleigh Taylor Wilkins, *The Problem of Burke's Political Philosophy*, Oxford: Clarendon Press, 1967, p. 61.

[40] Fleck, *Genesis*, p. 9.

[41] Arnold Hauser, *The Sociology of Art*, transl. by Kenneth J. Northcott, London: Routledge & Kegan Paul, 1974, pp. 28, 30, 21.

[42] Halbwachs, *Gedächtnis*, p. 389.

[43] Burton M. Leiser, *Custom, Law, and Morality: Conflict and Continuity in Social Behavior*, Garden City, N.Y.: Anchor Books, 1969, pp. 7–47.

[44] Cf. e.g. Geoffrey Sampson, *Liberty and Language*, Oxford: Oxford University Press, 1979, pp. 7 and 105.

[45] Dan I. Slobin – Thomas G. Bever, "Children use canonical sentence schemas: A crosslinguistic study of word order and inflections", *Cognition* 12 (1982), see esp. pp. 229 and 253.

[46] See Oakeshott's *Rationalism in Politics*, and Popper's paper referred to in note 27.

[47] See esp. the Lakatos-Musgrave and Gutting volumes referred to in notes 6 and 7, as well as Kuhn's collection *The Essential Tension: Selected Studies in Scientific Tradition and Change*, Chicago: The University of Chicago Press, 1977.

[48] David Hollinger, "T. S. Kuhn's Theory of Science and Its Implications for History", in: Gutting, ed., *Paradigms and Revolutions*, pp. 196 ff.

[49] *Ibid.*, pp. 197 f.

[50] "Comment on the Relations of Science and Art", in: T. S. Kuhn, *The Essential Tension*, pp. 346 and 348.

[51] *Ibid.*, p. 349.

[52] *Ibid.*

[53] *Ibid.*, p. 351.

[54] *The Sociology of Art*, pp. 30 f.

[55] Wittgenstein, *Lectures and Conversations on Aesthetics, Psychology and Religious Belief*, ed. by Cyril Barrett, Berkeley: University of California Press, 1967, p. 6.

[56] There are concepts, Musil writes, which for the poet constitute "his inherited tradition, concepts with whose help he has painstakingly consolidated his personal self. He does not even need to be in agreement with them all, he can strive to change them, yet he will still remain tied to them all much more than he is tied to the ground on which he walks. The poet is not only the expression of a momentary state of his soul – even should it be one that will introduce a new epoch. What he hands down is not decades but millenia old" ("Der Dichter in dieser Zeit", 1934, in: Robert Musil, *Gesammelte Werke*, ed. by Adolf Frisé, Reinbek bei Hamburg: Rowohlt, 1978, vol. 8, p. 1250). Or, as he put in the essay "Literat und Literatur" (1931): "Even the most independent writer does not produce anything which could not be shown to be almost without remainder dependent upon what has been handed down, both in form and content... One can only speak of originality where there is a tradition also" (*ibid.*, p. 1207). The connection between creativity and underlying traditions is explored in greater detail in W. Grassl and B. Smith, "A Theory of Austria", in: J. C. Nyíri, ed., *The Tradition of Austrian Philosophy*, Wien: Hölder-Pichler-Tempsky, 1986.

[57] *The Sociology of Art*, pp. 31 and 21.

[58] Karl Popper, *Unended Quest: An Intellectual Autobiography*, rev. ed., Fontana/Collins, 1976, p. 58.

[59] Carl Menger, *Untersuchungen über die Methode der Socialwissenschaften* (1883), in *The Collected Works of Carl Menger*, ed. F. A. von Hayek, vol. 2, London: The London School of Economics, 1934, p. 173.

[60] *Ibid.*

[61] *Ibid.*, p. 91.

[62] F. A. von Hayek, *Law, Legislation and Liberty*, vol. 3, London: Routledge & Kegan Paul, 1979, p. 167.

[63] *Truth and Method*, New York: Crossroad, 1975, pp. 36 f.

[64] H. L. A. Hart, *The Concept of Law*, 2nd ed., Oxford: Clarendon Press, 1963, p. 90.

[65] See ch. III of his *The Structure of Scientific Revolutions*, and esp. the paper "The Essential Tension: Tradition and Innovation in Scientific Research" (1959), in *The Essential Tension: Selected Studies in Scientific Tradition and Change*.

[66] "The Essential Tension", pp. 226 ff.

[67] "Only a dictionary", wrote Wittgenstein in his Preface, "makes it possible to hold the student completely responsible for the spelling of what he has written because it furnishes him with reliable measures for finding and correcting his mistakes, provided he has a mind to do so" (Wittgenstein, *Wörterbuch für Volksschulen*, ed. by A.

Hübner *et al.*, Wien: Hölder-Pichler-Tempsky, 1977, p. XXXI).

[68] Cf. e.g. Neville Bennett, *Teaching Styles and Pupil Progress*, London: Open Books, 1976. Current West-German perceptions are especially instructive. There, in the early 1970s, it has been declared that "broadening of linguistic competence" should supplant "training in the norms of 'standard German'" in general and the "learning of orthography" in particular. The results, as the progressive weekly *Spiegel* tells us, are by now catastrophic: "Den Reformern ging es nicht nur darum, die unterschiedlichen Ausgangspositionen auszugleichen zwischen Schülern, die aus einfachem Hause kamen, und anderen, die einem sprachlich besser ausgestatteten Milieu entstammten – ein erstrebenswertes Ziel. Nicht nur der manchmal aberwitzige Drill sollte abgeschafft werden, der Stumpfsinn, der oft die Deutschstunde beherrschte. – Dem Entwurf war auch zu entnehmen, 'daß diese «Hochsprache» bislang stets eine Gruppensprache gewesen ist, die als verbindliche Sprache durchgesetzt und bei der Schichtung der Gesellschaft als Mittel zur Stabilisierung dieser Schichtung benutzt worden ist'. ... Zum höheren Bildungsgut", runs the *Spiegel*'s mocking quote, "zählt seit jenen Tagen die 'Erweiterung der Fähigkeit, sich in umgangssprachlicher Kommunikation als realer Sprecher-Hörer mit anderen zu verständigen'" (*Der Spiegel*, July 9, 1984). But the ability of young Germans to write correctly, to read, and indeed to express themselves, has deteriorated drastically. And what sort of democracy is this, asks the *Spiegel*, where citizens are not capable of articulating their views?

[69] "What makes an individual a member of society and gives him claims is that he obeys its rules", writes F.A. von Hayek. "Wholly contradictory views may give him rights in other societies but not in ours. For the science of anthropology all cultures or morals may be equally good (though I doubt that this is true), but we maintain our society by treating others as less so" (*Law, Legislation and Liberty*, vol. 3, London: Routledge & Kegan Paul, 1979, p. 172).

[70] Karl W. Deutsch, *Nationalism and Social Communication: An Inquiry into the Foundations of Nationality*, London: 1953, p. 5. – On some important material determinants giving rise to feelings of nationalism see also Ernest Gellner, "Nationalism", in his *Thought and Change*, London: Weidenfeld & Nicolson, 1964.

[71] *Culture and Value*, p. 50.

NOTES TO CHAPTER 6

[1] From *trado+-tio*, with *trado* stemming from *trans-+do*.

[2] Cf. e.g. August Deneffe S. J., *Der Traditionsbegriff*, Münster/Westf.: 1931, p. 5.

[3] *Ibid.*, pp. 50 f.

[4] In his *Dictionary of the English Language*, completed by 1755, here quoted after the 1832 edition.

[5] It will be instructive to consider, briefly, the German usage of "Tradition", The Langenscheidt German-English Dictionary (1974 edition) gives three meanings: *Überlieferung*, translated as "tradition", and illustrated by phrases like "to cultivate a tradition", "to adhere to tradition", "to follow a tradition", "to break with tradition"; *Gepflogenheit*, translated as "convention, tradition", illustrated by "convention prescribes that"; and *Gewohnheit*, *Brauch*, translated by "tradition, (old) custom". The

terms "Überlieferung" and "Tradition" are explained in great detail by the *Deutsches Wörterbuch* of the Grimm brothers. "Überlieferung", we learn here, originally means *delivery*, but it has also acquired a more technical meaning: "auch verengt für *mündliche* mittheilung von geschlecht zu geschlecht, entsprechend dem theol. begriffe der tradition, der *schriftlichen* entgegengesetzt", i.e. there is an emphasis on *oral* delivery, and on the continuity between generations. On the other hand Goethe is cited as using the phrase "von gedruckten überlieferungen", *printed* traditions. And there is another quote from Goethe: "indem wir nun von überlieferung sprechen, sind wir unmittelbar aufgefordert, zugleich von autorität zu reden, denn genau betrachtet, so ist jede autorität eine art überlieferung", i.e. all authority is a kind of tradition. – The *word* "Tradition" did not exist in German before the 16th century; and until the 18th its use was almost exclusively confined to the religious sphere: "im kampf der reformation gegen die katholische kirche ins deutsche aufgenommen; anfänglich meistens von der einzelnen religiösen vorschrift oder einrichtung, die in confessioneller polemik als zusatz zu dem göttlichen gebot hingestellt wird". And a typical line by Lessing is quoted: "du [Luther] hast uns von dem joche der tr. erlöset", i.e. *thou, Luther, hath redeemed us from the yoke of tradition*. From the mid-18th century onwards the term then acquired its secular meaning of "mündlich und schriftlich überlieferte kunde von geschichtlichen begebenheiten", and by the 19th century the broader sense "das herkömmliche in haltung und handlung, das sich in socialen und geistigen gemeinschaften, in culturellen überlieferungszusammenhängen aller art fortpflanzt". The Grimm makes it clear that this broader sense, too, had a partly negative connotation: "ich will ... nicht mehr ruhen, bis mir nichts mehr wort und tr., sondern lebendiger begriff ist", wrote Goethe, and "ich will brechen mit tr. und konvention, will rang und namen von mir werfen und mit dem mann meiner wahl in süszer verborgenheit leben", wrote R. Baumbach – tradition and convention were seen as suppressing spontaneity in thought and in life.

6 He does mention it, e.g., in the first volume of his *Menschliches, Allzumenschliches: Ein Buch für freie Geister*. "Wer vom Herkömmlichen abweicht", he writes here, "ist das Opfer des Aussergewöhnlichen; wer im Herkömmlichen bleibt, ist der Sclave desselben. Zu Grunde gerichtet wird man auf jeden Fall" – he who stays within the boundaries of traditions is their slave (§ 552, quoted after the 1923 Musarion edition, vol. 8, p. 361).

7 Thus in the oft-quoted introductory passages of his third *Untimely Meditation*, "Schopenhauer as educator" (1874), according to which people are typically "timid", hiding themselves behind "customs and opinions", fearing their neighbours, who demand "conventionality", "Artists alone", writes Nietzsche, "hate this sluggish promenading in borrowed fashions and appropriated opinions and they reveal everyone's secret bad conscience, the law that every man is a unique miracle... The man who does not wish to belong to the mass needs only to cease taking himself easily; let him follow his conscience, which calls to him: 'Be your self! All you are now doing, thinking, desiring, is not yourself'. – Every youthful soul hears this call day and night and trembles when he hears it; for the idea of its liberation gives it a presentiment of the measure of happiness allotted it from all eternity – a happiness to which it can by no means attain so long as it lies fettered by the chains of fear and convention". Our era, Nietzsche goes on to write, is "ruled, not by living man, but by

pseudo-men dominated by public opinion ["öffentlich meinende Scheinmenschen"]; for which reason our age may be to some distant posterity the darkest and least known, because least human, portion of human history. ... It is so parochial to bind oneself to views which are no longer binding even a couple of hundred miles away. ... I will make an attempt to attain freedom, the youthful soul says to itself; and is it to be hindered in this by the fact that two nations happen to hate and fight one another, or that two continents are separated by an ocean, or that all around it a religion is taught which did not yet exist a couple of thousand years ago". (Friedrich Nietzsche, *Untimely Meditations*, transl. by R. J. Hollingdale, Cambridge: Cambridge University Press, 1983, pp. 127–129.)

[8] "Tradition and Some Other Forms of Order", *Proceedings of the Aristotelian Society*, N.S., vol. LIII (1953), pp. 2 and 8 f.

[9] "Tradition", in *Encyclopaedia of the Social Sciences*, New York: Macmillan, 1935, vol. 15, p. 62.

[10] "The Nature of Tradition", in D. M. Armstrong, *The Nature of Mind*, Harvester Press, 1981, pp. 89 and 100.

[11] Or *mediating fields* of particular actions, "Zwischenfelder, Zwischeninstanzen und Zwischenformen", as Waldenfels refers to them. He lists "Schema, Stil, Ritual, Symbol, Technik, Regel und Normen" (Bernhard Waldenfels, "Die Herkunft der Normen aus der Lebenswelt", in his *In den Netzen der Lebenswelt*, Frankfurt/M.: Suhrkamp, 1985, p. 134).

[12] This seems to be the meaning explicated by David K. Lewis, *Convention: A Philosophical Study*, Cambr., Mass.: Harvard Univ. Press, 1969, cf. esp. p. 42.

[13] For "Conventional" the *OED* lists, *inter alia*: "Relating to convention or general agreement; established by social convention; having its origin or sanction merely in an artificial convention of any kind; arbitrarily or artificially determined", "Characterized by convention; in accordance with accepted artificial standards of conduct or taste; not natural, original, or spontaneous", and "*Art.* Consisting in, or resulting from, an artificial treatment of natural objects; following accepted models or traditions instead of directly imitating nature or working out original ideas",

[14] Garden City, N.Y.: Anchor Books, 1969, pp. 7–47, cf. p. 56 above.

[15] Ruth Benedict, *Patterns of Culture*, Boston: 1934, p. 55.

[16] *Ibid.*, pp. 2 f.

[17] An aspect singled out, incidentally, also by Emile Durkheim. As Durkheim writes, "l'habitude n'est qu'une tendance à répéter automatiquement un acte ou une idée, toutes les fois que les mêmes circonstances la réveillent; elle n'implique pas que l'idée ou l'acte soient constitués à l'état de types exemplaires, proposés ou imposés à l'esprit ou à la volonté. C'est seulement quand un type de ce genre est préétabli, c'est-à-dire quand une règle, une norme est instituée, que l'action sociale peut et droit être présumée". (Durkheim, *Les formes élémentaires de la vie religieuse*, Paris: 1912, p. 620.) This might be the place to refer to Max Weber's corresponding definitions, all of which are, of course, coloured by his notion of "meaningful action" (*sinnhaftes Handeln*). "If an orientation toward social action occurs regularly", writes Weber, "it will be called 'usage' (*Brauch*) insofar as the probability of its existence within a group is based on nothing but actual practice. A usage will be called a 'custom' (*Sitte*) if the practice is based upon long standing. ... Usage also includes 'fashion' (*Mode*).

As distinguished from custom and in direct contrast to it, usage will be called fashion so far as the mere fact of the *novelty* of the corresponding behavior is the basis of the orientation of action. ... As distinguished from both 'convention' and 'law', 'custom' refers to rules devoid of any external sanction. The actor conforms with them of his own free will, whether his motivation lies in the fact that he merely fails to think about it, that it is more comfortable to conform, or whatever else the reason may be. For the same reasons he can consider it likely that other members of the group will adhere to a custom. – This custom is not 'valid' in anything like the legal sense; conformity with it is not 'demanded' by anybody. Naturally, the transition from this to validly enforced convention and to law is gradual. Everywhere what has been traditionally handed down [*das tatsächlich Hergebrachte*] has been the source of what has come to be valid". (Max Weber, *Economy and Society*, New York: Bedminster Press, 1968, p. 29. I have changed the last words of the translation.)

[18] New York: The Modern Library, pp. 70 and VI f.

[19] Johnson lists the meanings "Act of establishing", "Establishment; settlement", "Positive law", "Education". For the noun "institute" he gives "Established law; settled order" as the first set of meanings – illustrated by a qoute from Dryden: "This law, though custom now directs the course, / As nature's institute, is yet in force, / Uncancel'd, though disused" – and "Precept; maxim; principle" as the second. The Webster has "The act of instituting", like "Establishment; foundation; enactment" and "Instruction; education"; then "That which is instituted or established; as, (*a*.) Established order, or method, or custom; enactment; ordinance; permanent form of law or polity. 'The nature of our people, our city's *institutions*'. Shak. (b.) An established or organized society; an establishment, especially of a public character, or affecting a community; a foundation"; and finally "That which institutes or instructs; a treatise or textbook; a system of elements or rules". The *OED* renders, besides obsolete and/or technical senses like "The giving of form or order to a thing; orderly arrangement; regulation", "The established order by which anything is regulated", or "*Roman Law*. The appointment of a heir", the following major definition: "An established law, custom, usage, practice, organization, or other element in the political or social life of a people; a regulative principle or convention subservient to the needs of an organized community or the general ends of civilization".

[20] *Human Nature...*, p. 102.

[21] E. Shils, "Intellectuals, Tradition, and the Traditions of Intellectuals: Some Preliminary Considerations", *Daedalus*, Spring 1972, pp. 24 f. and 27.

[22] Bernhard Waldenfels, "The Ruled and the Unruly: Functions and Limits of Institutional Regulations", *Graduate Faculty Philosophy Journal* vol. 9, no. 1 (Winter 1982), p. 128.

[23] Pocock, "Time, Institutions and Action: An Essay on Traditions and Their Understanding", in: Preston King and B. C. Parekh, eds., *Politics and Experience*, Cambridge: Cambridge University Press, 1968, p. 215.

[24] *Op. cit.*, p. 63.

[25] As Allport writes: "To the ancients, *praejudicium* meant a *precedent* – a judgment based on previous decisions and experiences. ... Later, the the term, in English, acquired the meaning of a judgment formed before due examination and consideration of the facts... Finally the term acquired also its present emotional flavor of favorable-

ness or unfavorableness that accompanies such a prior and unsupported judgment. – Perhaps the briefest of all definitions of prejudice is: *thinking ill of others without sufficient warrant*". (Gordon W. Allport, *The Nature of Prejudice*, Garden City, N.Y.: Doubleday, 1958, p. 7.)

[26] The *Shorter Oxford English Dictionary* (1973 ed.) lists first "Injury, detriment, or damage, caused to a person by judgement or action in which his rights are disregarded; hence, injury to a person or thing likely to be the consequence of some action", and secondly "[*obs.*] A previous judgement; *esp.* a premature or hasty judgement [not recorded after:] 1835", "Preconceived opinion; bias favourable or unfavourable; prepossession; when used *absol.*, usu. with unfavourable connotation 1643", "an unreasoning predilection or objection 1654", and "[*obs.*] A preliminary or anticipatory judgement; an anticipation [not recorded after:] 1771".

[27] Burke, "Reflections on the Revolution in France" (1790), in: Burke, *The Works: Twelwe Volumes in Six*, 1887, Vol. III, pp. 346 f. Burke's conception of prejudice has been recently revived by Gadamer, who in his *Truth and Method* stresses that the "grundlegende Vorurteil der Aufklärung ist das Vorurteil gegen die Vorurteile überhaupt und damit die Entmachtung der Überlieferung. ... Das deutsche Wort Vorurteil scheint – wie das französische préjugé, aber noch entschiedener – durch die Aufklärung und ihre Religionskritik auf die Bedeutung 'unbegründetes Urteil' beschränkt worden zu sein" (Hans-Georg Gadamer, *Wahrheit und Methode*, 3rd ed., Tübingen: 1972, p. 255).

[28] Ernst Mach, *Popular Scientific Lectures* (1895), Fifth Edition, La Salle, Illinois: Open Court, 1943, pp. 214, 232 and 227.

[29] *The Man Without Qualities* (1930), transl. by E. Wilkins and E. Kaiser, London: Panther Books, 1968, vol. I, p. 52.

[30] Other senses listed by the Webster are: "A straight piece of wood, metal, or the like, which serves as a guide in drawing a straight line, or as a measuring rod for short distances", "The usual or appointed course of procedure; hence, behavior", "The administration of law; government; sway; empire; authority; control", "(*Law.*) An order regulating the practice of the courts, or an order made between parties to an action or suit", "(*Math.*) A determinate method prescribed for performing any operation and producing a certain result", and "(*Gram.*) A general principle concerning the formation or use of words, or a concise statement thereof".

[31] Emile Durkheim, *The Division of Labor in Society* (1893), transl. by G. Simpson, New York: The Free Press, 1964, p. 4.

[32] Oxford: Clarendon Press, 1961, pp. 8, 83, 54.

[33] Halbwachs, *Les cadres sociaux de la mémoire*, Paris: 1925, p. 295.

[34] Cf. chapter 5, pp. 55, above.

[35] Cf. "A Philosophical Inquiry into the Origin of our Ideas of the Sublime and Beautiful; with an Introductory Discourse concerning Taste", in: Edmund Burke, *The Works*, London: 1887, vol. 1, pp. 83 ff.

[36] As Gadamer puts it: "A phenomenon closely connected with taste is fashion. Here the element of social generalisation that the idea of taste contains becomes a determinig reality". Whereas, however, fashion is "a changeable law", taste represents "a constant whole of a social attitude". Gadamer stresses that the "history of the idea of taste" has a markedly political dimension in that it is "closely bound up with the

antecedents of the third estate. Taste is not only the ideal created by a new society, but we see this ideal of 'good taste' producing what was subsequently called 'good society'. Its criteria are no longer birth and rank but simply the shared nature of its judgments... Thus taste, in its essential nature, is not private, but a social phenomenon of the first order" (Hans-Georg Gadamer, *Truth and Method*, New York: 1985, pp. 34 f.). A related point is made by Wittgenstein when he remarks that taste is not a "mental state". We do not use the word "taste", he writes, as "the name of a feeling" (Wittgenstein, *The Blue and Brown Books*, Oxford: Basil Blackwell, 1958, pp. 143 f.).

[37] Where one could say that prejudice is facility in judgment, and skill is facility in performance, acquired through habit.

[38] *Human Nature and Conduct*, pp. 30, 64, 72, 184, 186.

[39] As Weber puts it, "purely traditional behavior" is marginal between "meaningful action" and "merely reactive behavior" (*op. cit.*, p. 4 f.). Weber intends to speak of "tradition" whenever the "regularities of action have become conventionalized", i.e., whenever "a statistically frequent action (*Massenhandeln*) has become a consensually oriented acation (*Einverständnishandeln*)" (*ibid.*, pp. 326 f.).

[40] Deneffe quotes an argument to the effect that "die kirchliche Tradition" will become a "nach eigenem Gutdünken zugestutzte", if it is located merely "*in den stummen Dokumenten* einer fernen Vergangenheit" (*op. cit.*, p. 124). Although emphasis on the exclusive authority of a written text does indeed involve a conservatism of its own (cf. e.g. the contrast Kümmel draws, in his "Jesus und der jüdische Traditionsgedanke", between the aristocratic Sadducees, denying the obligation of the unwritten law, and the Pharisees, holding "den revolutionären Gedanken einer fortschreitenden mündlichen Tradition", in: W. G. Kümmel, *Heilsgeschehen und Geschichte*, Marburg: 1965, p. 25), such a text still offers, epistemologically speaking, the freedom of reflection. The two aspects are plausibly brought together by Ferdinand Hahn. "Mündliche Tradition", Hahn writes, "hat ihre eigenen Gesetze und ihre eigene Ausprägung. Mündliche Tradition lebt mit und lebt von der Traditionsweitergabe; sie läßt sich vom Akt der Tradierung nicht trennen, und diese Tradierung ist kein abgeschlossener Prozeß. Schriftliche Tradition dagegen tendiert auf Abschluß. Ihr geht es um Sammlung, Zusammenfassung und Vereinheitlichung. An die Stelle des Tradenten tritt fortan der Interpret. ... Beim Übergang von der mündlichen Überlieferung zur schriftlichen Gestaltung tritt ein deutlicher Funktionswechsel ein... Der Text gewinnt eine selbständige Stellung. Die ursprüngliche Unmittelbarkeit des aktuellen Bezuges zwischen dem Sprechenden und dem Hörenden besteht nicht mehr, der Text tritt gleichsam zwischen den Tradenten und den Hörer". (F. Hahn, "Zur Verschriftlichung mündlicher Tradition in der Bibel", *Zeitschrift für Religions- und Geistesgeschichte* 39, 1987, pp. 314 and 316.)

[41] F. H. Levi, *An Introduction to Legal Reasoning*, 4th ed., University of Chicago Press, 1955, p. 41, quoted by Bruno Leoni, *Freedom and the Law*, Princeton, N.J.: Van Nostrand, 1961, p. 186.

[42] *Op. cit.*, pp. 66 and 63. "Customs", writes Radin, "are matters of fact; they exist only as long as they are widely practised and generally recognized, and when a custom is merely a memory it ceases to be a custom. But a tradition may well be merely a memory, retained by an extremely small number of persons. In fact it is generally when the tradition is no longer a description of actual fact and when it has

become somewhat evanescent as a rule of conduct that it most clearly justifies its name and performs its real functions" (*ibid.*, p. 66).

[43] *Op. cit.*, p. 295.

[44] Josef Pieper, "Über den Begriff der Tradition", *Arbeitsgemeinschaft für Forschung des Landes Nordrhein-Westfalen: Geisteswissenschften*, Heft 72, Köln: Westdeutscher Verlag, 1958, p. 20. Pieper refers to Augustine, for whom *memoria* was "etwas Überindividuelles, eine Kraft, die über die Generationsfolge hinwegzudringen und Erfahrungen zurückzurufen vermöge, die dem Menschen in der Ursprungsfrühe seiner Geschichte zuteil geworden sind" (*ibid.*, p. 32).

[45] "Traditions are beliefs, standards, and rules, of varying but never exhaustive explicitness, which have been received from the preceding generation, through a process of continuous transmission from generation to generation. ... Traditions possess authority by virtue of the quality which they acquire in the minds of persons of one generation when they believe these traditions were accepted by a succession of ancestors coming up to the immediate past. – The traditional rule possesses authority because its acceptance establishes an attachment to the past of a family, town, country or corporate body to which an inherent value is attributed. ... Acceptance of tradition is the creation of a state of communion with past powers: It is of the same order as any act of communion with one's contemporary society... [S]tability through generations of a belief or practice does not constitute tradition... Reception must be accompanied by affirmative attachment to the past, however vague, unconscious, and unspoken. The performance of an action which is presented from the past by authority but which is performed only because no other alternative mode of action can be imagined is at the margin of tradition. ... Traditional attachment implies receptive affirmation, neither reception without affirmation nor affirmation without reception". ("Tradition and Liberty: Antinomy and Interdependence", *Ethics* LXVIII/3, April 1958, pp. 154 f.) In his book *Tradition* (London: Faber and Faber, 1981) Shils stresses the technical terms "substantive tradition" ("traditions which maintain the received", p. 4) and "substantive traditionality" ("the appreciation of the accomplishments and wisdom of the past and of the institutions especially impregnated with with tradition", p. 21), retaining, for the term "tradition" itself, the widest possible meaning: "Tradition means many things. In its barest, most elementary sense, it means simply a traditum; it is anything which is transmitted or handed down from the past to the present. It makes no statement about what is handed down or in what particular combination or whether it is a physical object or a cultural construction; it says nothing about how long it has been handed down or in what manner... The degree of rational deliberation which has entered into its creation, presentation, and reception likewise has nothing to do with whether it is a tradition. ... Tradition – that which is handed down – includes material objects, beliefs about all sorts of things, images of persons and events, practices and institutions. It includes buildings, monuments, landscapes, sculptures, paintings, books, tools, machines" (*ibid.*, p. 12).

[46] Tradition, Pocock writes, is "the handing on of formed ways of acting, a formed way of living, to those beginning or developing their social membership... A tradition, in its simplest form, may be thought of as an indefinite series of repetitions of an action, which on each occasion is performed on the assumption that it has been performed before; its performance is authorised – though the nature of authorisation

may vary widely – by the knowledge, or the assumption, of previous performance. In the pure state, as it were, such a tradition is without a conceivable beginning; each performance presupposes a previous performance, in infinite regress. Furthermore, it may well be that it is the assumption, rather than the factual information, of previous performance that is operative" (*op. cit.*, pp. 209 and 212). That the factual information conveyed by a tradition might be spurious and the tradition still be functionally effective, was made explicit in the term *invented traditions*, coined by Eric Hobsbawm. These are taken to mean "a set of practices, normally governed by overtly or tacitly accepted rules and of a ritual or symbolic nature, which seek to inculcate certain values and norms of behaviour by repetition, which automatically implies continuity with the past". (Hobsbawm, "Introduction: Inventing Traditions", in: Hobsbawm and Ranger, eds., *The Invention of Tradition*, Cambridge: Cambridge Univ. Press, 1983, p. 1.)

[47] *Op. cit.*, pp. 100 f. – Armstrong observes that "[t]raditions are *unitary* things in a way that mere customs are not", i.e. similar traditions in different societies count as *different* traditions, while similar customs in different societies can still be said to be just the same ones (*ibid.*). This, too, seems to be a corollary of the fact that specific traditions are *known*, or are *supposed*, to have been handed down in specific, unique, circumstances.

[48] These generations, as for instance H. B. Acton pointed out, "need not be biological ones; traditions exist, for example, in a school, if a belief or practice extends unquestioned over several successions of student careers" (*op. cit.*, p. 3).

[49] The concept we adopt is, then, rather more specific than that taken for granted by the influential German author Hermann Lübbe, who defines traditions as "erfahrungsbewährte Lebensformen..., die kraft generationsüberdauernder Geltung schließlich den Status kultureller Selbstverständlichkeiten gewinnen" (Lübbe, "'Neo-Konservative' in der Kritik", *Merkur* 1983/6, p. 624), or simply as "stabile praxis-leitende Orientierungssysteme" (Lübbe, "Rationalitätsverluste. Über Tendenzen der Flucht aus der Gegenwart", *Jahrbuch 1982 der Technischen Universität München*, p. 100).

[50] An obvious minimum requirement, as suggested e.g. by 1 Cor 15, 3, or indeed by Augustine's words "quod a patribus acceperunt, hoc filiis tradiderunt" (*Contra Iulianum*, 2, 10, 34).

[51] Especially since the element of recollection, as Halbwachs has demonstrated, invariably involves reconstruction, and indeed *construction*: "L'opération de la mémoire suppose en effet une activité à la fois constructive et rationelle de l'esprit... elle ne s'exerce que dans un milieu naturel et social ordonné, cohérent... Tout souvenir, si personnel soit-il ... est en rapport avec tout un ensemble de notions que beaucoup d'autres que nous possèdent, avec des personnes, des groupes, des lieux, des dates, des mots et formes du langage, avec des raisonnements aussi et des idées, c'est-à-dire avec toute la vie matérielle et morale des sociétés dont nous faisons ou dont nous avons fait partie" (*op. cit.*, p. 38).

NOTES TO CHAPTER 7

[1] Durkheim wrote in this connection that there was no society which did not feel the need "from time to time to relive its great moments in thought", "to fix memories in celebration" and thereby to strengthen and to bring to life "collective feelings" and "collective ideas". (*Les formes élémentaires de la vie religieuse*, Paris: 1912, pp. 610 f.) In this sense Durkheim's disciple Maurice Halbwachs asks how "a society of whatever sort could exist, survive and become conscious of itself if it did not embrace in one view a totality of present and past events, if it did not have the possibility of tracing back through time and of constantly moving back and forth over the traces it has left". (*La mémoire collective*, 1950, 2nd ed., Paris: 1968, p. 129.) Those "who demand from memory only the clarification of their own immediate actions, and for whom the past ... would have the same colouring as the present, would have not the slightest sense of social continuity". (*Les cadres sociaux de la mémoire*, Paris: 1925, p. 113.) In this sense Edward Shils indeed identifies the attachment to the past of a group with the attachment to the group itself. ("Acceptance of tradition is the creation of a state of communion with past powers: It is of the same order as any act of communion with one's contemporary society", Shils, "Tradition and Liberty: Antinomy and Interdependence", *Ethics* LXVIII/3, 1958, pp. 154 f.) And in this sense Ong says: "If man is to retain sanity through affirmation of his past, as he must, certain times and places ... will retain to a degree a special sacrality", to which he however adds the remark, crucial in the context of the present argument: "But time and place as foci of the human life-world have been complemented and in part supplanted by the sense of conscious interchange of man with man, a sense of human presence, of intersubjectivity, of psychic participation even with persons in the most distant lands", Walter J. Ong, SJ, *The Presence of the Word: Some Prolegomena for Cultural and Religious History*. New Haven: Yale University Press, 1967, pp. 311 f.

[2] Thus it by no means signifies an "undermining of social bonds" if the individual appears merely as a node in the "circulation of information" – if, as Lyotard suggests, society appears to us as a "system in the cybernetic sense", as a mere "network of communication, with intersections" (Jean-François Lyotard, *La condition postmoderne*, Paris: 1979, pp. 31 f.). "Communication creates, or makes possible at least", as Robert E. Park expressed it already in 1938, "that consensus and understanding among the individual components of a social group which eventually gives it and them the character not merely of society but of a cultural unit. It spins a web of custom and mutual expectation which binds together social entities... Family group or labor organization, every form of society except the most transient has a life-history and a tradition. It is by communication that this tradition is transmitted. It is in this way that the continuity of common enterprises and social institutions is maintained, not merely from day to day, but from generation to generation" (Park, "Reflections on Communication and Culture", *The American Journal of Sociology*, Sept. 1938, pp. 191 f.). The vague distinction Park here draws between "culture" and "society" will in the subsequent literature again recede into the background. Thus Moss (actually referring to Dewey) could write that "society exists in and through communication; common perspectives – common cultures – emerge through participation in common communication channels" (Gordon E. Moss, "Identification and the Curve of Optimal

Cohesion", in: George G. Haydu, *Experience Forms: Their Cultural and Individual Place and Function*, The Hague: Mouton, 1979, p. 210). Not so Karl W. Deutsch, who in his *Nationalism and Social Communication* makes following distinction: "a community consists of people who have learned to communicate with each other... Societies produce, select, and channel goods and services. *Cultures produce, select, and channel information*. – There is no community nor culture without society. And there can be no society, no division of labour, without a minimum of transfer of information, without communication. Yet the difference between society and community is crucial" (London – New York: 1953, pp. 65 f. and 69). More immediately relevant to our present problem, Deutsch makes the observation (originating, in the last analysis, with Halbwachs) that *the past is but an implication of the communicational situation here and now*: "the relatively coherent and stable structure of memories, habits, and values ... depends on existing facilities for social communication, both from the past to the present and between contemporaries. ... Instead of being automatically united by a shared history, men at least under some conditions cannot share the historical events through which they live, unless they are already in some sense united" (*ibid.*, pp. 49 and 5).

3 I follow here above all the work of Walter J. Ong: *Orality and Literacy: The Technologizing of the Word*, London: Methuen, 1982. Cf. also the works of Eric Havelock: *Preface to Plato*, Cambridge, Mass.: Harvard University Press, 1963; *The Greek Concept of Justice: From Its Shadow in Homer to Its Substance in Plato*, Cambridge, Mass.: Harvard University Press, 1978; *The Literate Revolution in Greece and Its Cultural Consequences*, Princeton University Press, 1982; *The Muse Learns to Write: Reflections on Orality and Literacy from Antiquity to the Present*, New Haven: Yale University Press, 1986; as well as Jack Goody and Ian Watt, "The Consequences of Literacy", 1963, in: Goody (ed.), *Literacy in Traditional* Societies, Cambridge: Cambridge Univ. Press, 1968.

4 As Roy A. Rappaport writes: "there is likely to be a connection between sanctity and the special characteristics of human communication... It is possible to argue that the survival of any population depends upon social interactions characterized by some minimum degree of orderliness and that orderliness depends upon communication. But communication is effective only if the recipients of messages are willing to accept, as being in at least some minimum degree reliable, the messages which they receive" ("Ritual, Sanctity, and Cybernetics", *American Anthropologist* 73, 1971, pp. 67 f.). See also Wolfgang Rösler: "The 'oral poet' unveils his conception in the call of the muses... ... The poet – it becomes clear – sees himself as medium of divine knowledge; what he sings has occurred, is true. This claim in never retracted, relativized or made problematic" (Rösler, "Schriftkultur und Fiktionalität. Zum Funktionswandel der griechischen Literatur von Homer bis Aristoteles", in: A. u. J. Assmann – Chr. Hardmeier, eds., *Schrift und Gedächtnis: Archäologie der literarischen Kommunikation I*, München: 1983, p. 110).

5 This is of course the great theme and at the same time the unsolved problem of Innis. "An oral tradition implies freshness and elasticity but students of anthropology have pointed to the binding character of custom in primitive cultures" (Harold A. Innis, *The Bias of Commmunication*, University of Toronto Press, 1951, repr. 1971, p. 4). Cf. Rappaport: "To the extent that the discourse of religion, religious ritual and

religious experience contribute to the maintenance of orderliness and the reduction of anxiety without contributing to the correction of the factors producing the anxiety and disorder they are not adaptive but pathological", *loc.cit.*, p. 73.

6 As Goody and Watt, building here mainly on Halbwachs, write: "The social function of memory – and of forgetting – can thus be seen as the final stage of what may be called the homeostatic organization of the cultural tradition in non-literate society". ("Consequences of Literacy", in: Goody, ed., p. 30.)

7 *The Muse Learns to Write*, p. 58.

8 Goody, ed., p. 34. B. A. van Groningen expressed this same idea in his book *In the Grip of the Past: Essay on an Aspect of Greek Thought*. Van Groningen does not, however, make the orality/literacy distinction. "The Greeks often refer to the past and, by doing so, they bind the matter in question to a chronological conception. But as soon as we inquire after the real meaning, it becomes obvious that the idea is not temporal but is used in a general sense" (Leiden: E. J. Brill, 1953, p. 17). Cf. also M. I. Finley, "Myth, Memory, and History", *History and Theory* 1965, pp. 284 f.: Homeric "epic was *not history*. ... Like all myth, it was timeless. Dates and a coherent dating scheme are as essential to history as exact measurement is to physics. Myth also presented concrete facts, but these facts were completely detached: they were linked neither with what went before nor with what came after".

9 A recent contribution to this literature is Barry Smith's paper "Textual Deference", *American Philosophical Quarterly* 28/1, 1991, pp. 1–13, which examines the role of the *commentary genre* in various philosophical schools of the past and present. The commentary is, of course, a creature of the transition to the phase of literacy (a transition which, as we shall see below, is *completed* only with the emergence of the printed book). For the commentary is in essence a product of reflection on a text treated as a fixed and sacred repository of wisdom.

10 *The Decline of the West*, New York: 1934, vol. II, pp. 149 f.

11 Koselleck, *Vergangene Zukunft: Zur Semantik geschichtlicher Zeiten*, Frankfurt/M.: Suhrkamp, 1979, p. 18.

12 As the Hungarian historian Hajnal, writing on the High Middle Ages, observes: "Though waxed tablets might have been widely employed in the course of quick composition and recording, the fact remains that the time-honoured methods of the education of the *clerici* centered around severe drilling via word-of-mouth... the text to be written must have taken definite and exact shape in the mind prior to its being 'copied' on parchment... It is well known how teaching at the universities proceeded without books and without writing: at the *lectio publica* a strictly compulsory traditional book in the teacher's hand; there is lecturing, detailed explanation, repeated over and over again... But the students themselves at their *hospitia* are preparing in advance for the text of the daily lecture, their masters and seniors reciting it loudly into their ears, and as soon as the lecture is over, they repeat the text again and again. ... It is simply indispensable for a student to have groups of mates, and elders around him; they are his living educational tools, carriers of scientific material available for exercises". (Hajnal, "Universities and the Development of Writing in the XIIth-XIIIth Centuries", *Scriptorum. International Review of Manuscript Studies*, VI/2, 1952, pp. 179 f. Published translation slightly amended.) The emergence of printing did not, of course, change this cultural pattern in the twinkling of an eye.

[13] It is significant that Koselleck, while conceiving the Reformation as the decisive factor in the emergence of the modern historical consciousness, pays no attention at all to the printing of books, in spite of the crucial contribution of the latter to the spreading of Reformation ideas.

[14] *The Printing Press as an Agent of Change: Communications and Cultural Transformations in Early-Modern Europe*, Cambridge: Cambridge University Press, 1979, vol. I, pp. 74 f. and 124. It was, according to Ong, as a result of typography that philology came into being: "The new attention to textual accuracy in documents coming out of the remote past generated a consciousness of the differences between past and present in man's life-world such as had never before been known", *The Presence of the Word*, p. 274.

[15] Eisenstein, *op. cit.*, p. 301.

[16] As Gadamer puts it, there is "a fundamental distance between the present and all that is historically preserved" (Hans-Georg Gadamer, *Wahrheit und Methode*, 2nd ed., Tübingen: 1965, p. XIX). In his *The Death of the Past* J. H. Plumb speaks of the ability "to see things as they were in their own time", "the consciousness of a different past", the "wish to understand the past in its own terms" (London: Macmillan, 1969, pp. 82 and 118 f.).

[17] Koselleck, *op. cit.*, p. 364.

[18] *Ibid.*, p. 366.

[19] *Ibid.*, pp. 366 f.

[20] *Ibid.*, pp. 359 and 368.

[21] Cf. Goody and Watt: "even within a literate culture, the oral tradition – the transmission of values and attitudes in face-to-face contact – nevertheless remains the primary mode of cultural orientation", *op. cit.*, p. 58.

[22] Hajnal, "A technika fejlödése", in: *Domanovszky Emlékkönyv*, 1937, pp. 229 f. In this context Hajnal speaks also of the "lively traditional social organization" of technically perfect early cultures, cf. *ibid.*, p. 238.

[23] Goody – Watt, *op. cit.*, pp. 88 f.

[24] *Ibid.*, p. 58. The reference is to the *Unzeitgemäße Betrachtung*, "Vom Nutzen und Nachteil der Historie für das Leben", cf. e.g. *Sämtliche Werke: Kritische Studienausgabe*, vol. 1, pp. 273 f., 267 f. and 250.

[25] Haefner, *Die neue Bildungskrise*, Reinbek bei Hamburg: Rowohlt, 1985, pp. 31 f.

[26] As Plumb puts it: "There are human truths to be derived from history... Some, if not the most important, of the problems which face society today are not new ones; there are similarities and analogies in the past. Any process which increases man's awareness of himself, that strengthens his chance of controlling himself and his environment, is well worth pursuing", *op. cit.*, p. 17.

[27] "Despite what is sometimes said, electronic devices are not eliminating printed books but are actually producing more of them. Electronically taped interviews produce 'talked' books and articles by the thousands" (Ong, *Orality and Literacy: The Technologizing of the Word*, p. 135).

[28] "[W]ith telephone, radio, television and various kinds of sound tape", writes Ong, "electronic technology has brought us into the age of 'secondary orality'. This new orality has striking resemblances to the old in its participatory mystique, its fostering of a communal sense, its concentration on the present moment, and even its use of

formulas... But it is essentially a more deliberate and self-conscious orality, based permanently on the use of writing and print, which are essential for the manufacture and operation of the equipment and for its use as well", *ibid.*, p. 136. "Turing's man", as J. David Bolter writes, "has in fact inherited traits from both the ancient and Western European characters, and the very combination of these traits makes him different from either" (Bolter, *Turing's Man: Western Culture in the Computer Age*, Chapel Hill: Univ. of North Carolina Press, 1984, p. 218).

29 Ong, *The Presence of the Word*, pp. 301 f. und 313.

30 Cf. e.g. *Zeit-Verhältnisse: Zur Kulturphilosophie des Fortschritts* (Graz – Wien – Köln: Styria, 1983).

31 Bolter, *op. cit.*, p. 225.

32 *Ibid.*, p. 123.

33 G. Richard Dimler, S. J., "Word Processing and the New Electronic Language", *Thought*, vol. 61, no. 243 (Dec. 1986), p. 463. I am indebted to Tibor Frank for having drawn my attention to this issue of *Thought*.

34 Cf. Robert Sokolowski, "Natural and Artificial Intelligence", *Daedalus*, Winter 1988, p. 49: "The gradual diffusion of writing into human affairs can serve as a historical analogue for the seepage of artificial intelligence into human exchanges. Writing did not simply replace the linguistic activities that people carried out before there was writing; its major impact was to make new kinds of activity possible and to give a new shape to old kinds. It enlarged and differentiated economic, legal, political, and aesthetic activities, and it made history possible. ... Writing did all this by amplifying intelligence. Printing accelerated the spread of the written word, but it did not change the nature of writing. – The question that can be put to artificial intelligence is whether it is merely an extension of printing or a readjustment in the human enterprise that began when writing entered into human affairs. Word processing is clearly just a refinement of printing, a kind of glorified typing, but artificial intelligence appears to be more than that. It seems to be able to reform the embodiment of thought that was achieved in and by writing".

35 Consider, as a typical example, the formulation of McLuhan: "As unfallen Adam in the Garden of Eden was appointed the task of contemplation and naming of creatures, so with automation. We have now only to name and program a process or a product in order for it to be accomplished. ... The entire world, past and present, now reveals itself to us like a growing plant in an enormously accelerated movie. Electric speed is synonymous with light and with the understanding of causes" (Marshall McLuhan, *Understanding Media: The Extensions of Man*, New York: McGraw-Hill, 1964, p. 352).

36 But see already Heidegger in a revealing passage in "The Anaximander Fragment": "All historiography predicts what is to come from images of the past determined by the present. It systematically destroys the future... Historicism has today not only not been overcome, but is only now entering the stage of its expansion and enttrenchment. The technical organization of communications throughout the world by radio and by a press already limping after it is the genuine form of historicism's dominion" (*Early Greek Thinking*, transl. by David Farrell Krell and Frank A. Capuzzi, San Francisco: Harper & Row, 1984, p. 17).

[37] Dijkstra, *Selected Writings on Computing: A Personal Perspective*, New York: Springer, 1982, p. 130.

[38] Mach, "On the Classics and the Sciences", in: *Popular Scientific Lectures*, La Salle, Ill.: Open Court, 1986, pp. 344 and 358.

[39] Haefner, *Die neue Bildungskrise*, pp. 191 ff.

NOTES TO CHAPTER 8

[1] Quoted from K. T. Fann, ed., *Wittgenstein: The Man and His Philosophy*, New York: Dell, 1967, p. 35.

[2] Cf. Charles F. Hockett, *The State of the Art*, The Hague: Mouton, 1968, p. 83.

[3] Einar Haugen, "Linguistics and Language Planning", in E. Haugen, *Studies*, ed. by E. Sch. Firchow *et al.*, The Hague: Mouton, 1972, p. 513.

[4] Cf. chapters 1 and 2 above.

[5] I try to show this in my paper "Wittgenstein and the Problem of Machine Consciousness", *Grazer Philosophische Studien*, vol. 33/34 (1989), pp. 375–394. Hans Sluga's highly interesting essay in the same volume, "Thinking as Writing", examines Wittgenstein's method of literary composition, without however noting its postmodern/post-literate aspects.

[6] The strongest argument against Esperanto as an ordinary language is, after all, that it *could not have been devised under conditions of orality*.

[7] Ludwig Wittgenstein, *Culture and Value*. Transl. by Peter Winch, Oxford: Basil Blackwell, 1980, p. 52e.

[8] R. A. Hall, *Linguistics and Your Language*, New York: Doubleday, 1960, p. 45. Hall's book was originally published in 1950, under the title *Leave Your Language Alone!*

[9] *Ibid.*, pp. 39 f.

[10] *Ibid.*, p. 41, cf. also p. 200.

[11] *Ibid.*, p. 40.

[12] *Ibid.*, pp. 201 ff. – Hall here is repeating (almost verbatim) some of Bloomfield's arguments in his classic work, *Language*. However, Bloomfield's stress was more on the primacy of "phonemic habits" than on the setbacks of there being no authoritarian means available by which radical innovations could be implemented. "We may expect", Bloomfield wrote, "that at some time in the future our social organism will reach a degree of co-ordination and flexibility where a concerted change becomes possible, or else that mechanical devices for reproducing speech will supersede our present habits of writing and printing" (Leonard Bloomfield, *Language* [New York: 1933] p. 503).

[13] Valter Tauli, "The Theory of Language Planning", in Joshua A. Fishman, ed., *Advances in Language Planning*, The Hague: Mouton, 1974, pp. 52 and 61.

[14] That this alteration, incidentally, might have amounted to a loss instead of a gain, was asserted emphatically by the writer and essayist László Németh, in his *Kisebbségben*, published in 1939.

[15] G. Bárczi, *A magyar nyelv életrajza*, Budapest: Gondolat, 1975, pp. 309 and 350.

[16] Y. Alloni-Fainberg, "Official Hebrew Terms for Parts of the Car", in Fishman, ed., *Advances...*, p. 495.

[17] Joseph H. Greenberg, *Essays in Linguistics*, New York: Wenner-Gren, 1957, p. 65.
[18] E. Haugen, "Instrumentalism in Language Planning", in Joan Rubin and Björn H. Jernudd, eds., *Can Language Be Planned?* (The University Press of Hawaii, 1971, pp. 286 f.)
[19] Alloni-Fainberg, *op. cit.*, p. 494.
[20] C. A. Ferguson, "Sociolinguistic Settings of Language Planning", in Joan Rubin *et al.*, eds., *Language Planning Processes*, The Hague: Mouton, 1977, pp. 15 ff.
[21] *Ibid.*, p. 14.
[22] Thus in R. A. Hall's book. According to Hall's eventually self-defeating argument, the issue of "correct" vs. "incorrect" speech forms "boils down, really, to a question of acceptability in certain classes of our society, in those classes which are socially dominant and which set the tone for others.... 'Correct' can only mean 'socially acceptable', and apart from this has no meaning as applied to language" (*op. cit.*, p. 13). Hall then goes on to point out that the "entire attempt to set up absolute standards, rigid norms, for regulating people's language is destined to failure from the outset, because ... 1) there is no authority that has either the right or the ability to govern people's usage; and 2) such an authority, even when it has been set officially (as were the French and Spanish Academies), can never find valid standards by which to govern usage" (*op. cit.*, p. 26). Still, abandonment of absolute standards, Hall reassures his readers, "does not necessarily mean abandonment of all standards". One should rely on "the cohesive, centripetal forces of society": "the pressure of human need for communication will always insure people's keeping their speech reasonably uniform" (*op. cit.*, p. 250). What Hall does not stress is that the centripetal forces he refers to essentially depend on the role of elites – of paradigmatic groups in social life, business, science, and technology.
[23] Cf. Haugen, "Linguistics and Language Planning", *loc. cit.*, p. 517.
[24] *Ibid.*, p. 518. – "The United States", Haugen adds, "declared its linguistic independence of England by substituting another private citizen, Noah Webster, for the English Johnson".
[25] *Ibid.*, p. 524.
[26] Cf. István Szerdahelyi, "Entwicklung des Zeichensystems einer internationalen Sprache: Esperanto", in István Fodor – Claude Hagège, eds., *Language Reform: History and Future*, Hamburg: Buske Verlag, 1983/84, vol. 3, pp. 287 f.
[27] Hall, *op. cit.*, p. 234.
[28] As Géza Bárczi wrote in the journal *Eszperantó Magazin* (Budapest: 1975, "Az eszperantó nyelvi értéke"): "Esperanto suffers from an illness: it ... is not carried by a homogeneous community which would uniformly reject any tampering with it", – In the natural sciences, where homogeneity within each field is characteristically maintained by the authority of scientific elites, the uniformity of specialists' universal languages (auxiliary notations) is more easily preserved. Another factor of course is that such languages have a relatively limited range of application. For a discussion of some pertinent problems see Elmar Holenstein, *Menschliches Selbstverständnis*, Frankfurt/M.: Suhrkamp, 1985, pp. 164 ff. Cf. also Bloomfield, *Language*, pp. 506 f.
[29] Cf. Szerdahelyi, *op. cit.*, pp. 294 and 299 f.
[30] K. Brugmann – August Leskien, *Zur Kritik der künstlichen Weltsprachen*, Straßburg: 1907, p. 24.

[31] For a thorough discussion of the issue see Wallace L. Chafe, "Idiomaticity as an Anomaly in the Chomskyan Paradigm", *Foundations of Language*, vol. 4 (1968), p. 111: "The importance of idioms in language", writes Chafe, "cannot be doubted. Their ubiquity makes them anything but a marginal phenomenon..."

[32] W. F. Twaddel, quoted by Dwight Bolinger, "Meaning and Memory", in George G. Haydu, ed., *Experience Forms: Their Cultural and Individual Place and Function*, The Hague: Mouton, 1979, p. 96.

[33] J. J. Katz – J. A. Fodor, "The Structure of a Semantic Theory", *Language*, vol. 39 (1963), p. 171.

[34] Cf. Hockett, *op. cit.*, pp. 44 ff., esp. p. 57. – The gist of these arguments is of course foreshadowed in the *Blue Book*: "remember that in general we don't use language according to strict rules... In practice we very rarely use language as ... a calculus" (Oxford: Basil Blackwell, 1958, p. 25).

[35] Dwight Bolinger, "The Atomization of Meaning", *Language*, vol. 41 (1965), p. 570. – This is not to claim that people cannot say new things. As Hockett writes: "One assumes that, in a situation in which various partly incompatible patterns are all apt, their interplay and the resolution of their incompatibilities can lead to a sentence that has not been said before. ... An individual's language, at any given moment, is a set of habits – that is, of analogies. ... In many, perhaps most, speech situations, many different analogies are at work" (Hockett, *op. cit.*, pp. 67 and 92 f.).

[36] Bloomfield, "Meaning", *Monatshefte für Deutschen Unterricht*, vol. 35 (1943), p. 104.

[37] K. W. L. Heyse, *System der Sprachwissenschaft*, Berlin: 1856, pp. 6 f.

[38] Cf. *Philosophical Investigations*, Part I, § 43.

[39] Cf. Bolinger, "The Atomization of Meaning", pp. 568 and 570.

[40] Thus the idea of the conventional, in the sense of *arbitrary*, nature of language has an only very limited validity. This was clearly seen by Humboldt. As O. Fr. Bollnow wrote in his "Wilhelm von Humboldts Sprachphilosophie": "Das Wort ist nicht ein beliebig vertauschbares Zeichen für einen ihm schon vorgegebenen reinen Begriff. So etwas ist nur in dem ganz engeren Bereich solcher Begriffe möglich, 'welche durch bloße Construction erzeugt werden können, oder sonst rein durch den Verstand gebildet sind' ([Humboldts Gesammelte Schriften] IV 21/22), also auf dem Gebiet der Naturwissenschaft, soweit diese in einen rein deduktiven Gang auflösbar ist. Wo wir uns dagegen auf dem Boden der gewachsenen Sprache", i.e. *organically grown language*, "befinden, vollzieht sich die Begriffsbildung erst in und mit dem Wort und läßt sich nicht davon lösen" (*Zeitschrift für deutsche Bildung*, vol. 15 [1938], p. 108).

[41] Hockett, *The State of the Art*, p. 68. – Through one's *mother tongue*, wrote Heyse echoing Humboldt, "sind wir ein Glied der natürlichen menschlichen Gesammtheit, zunächst der Familie, sodann der Nation, der wir angehören, und fühlen uns als ein solches" (Heyse, *op. cit.*, p. 2).

[42] Cf. Newton Garver, "Die Lebensform in Wittgensteins *Philosophischen Unter-suchungen*" and Rudolf Haller, "Lebensform oder Lebensformen?", both in *Grazer Philosophische Studien*, vol. 21, 1984, as well as Eike von Savigny, "Viele gemein-same menschliche Handlungsweisen", in: W. L. Gombocz *et al.*, eds., *Traditionen und Perspektiven der analytischen Philosophie: Festschrift für Rudolf Haller*, Wien: Hölder-Pichler-Tempsky, 1989, pp. 224 f.

NOTES TO CHAPTER 9

[1] "Wittgenstein und Heidegger. Die Frage nach dem Sinn von Sein und der Sinnlosigkeitsverdacht gegen alle Metaphysik" (1967), in: O. Pöggeler (ed.), *Heidegger: Perspektiven zur Deutung seines Werks*, Königstein/Ts.: Athenäum, 1984, p. 358.

[2] Reinhard Merkel, "Das Klappen der Schere des Haarschneiders". *Die Zeit*, Sept. 1, 1989, p. 45.

[3] George Steiner, *Martin Heidegger: Eine Einführung*, München: Carl Hanser Verlag, 1989, p. 17. The passage is from the introduction to the German edition.

[4] *Op. cit.*, p. 360.

[5] *Ibid.*, p. 367.

[6] *Ibid.*, pp. 375 f. – One should add here that thereby the aspect of *silence* which stands in the foreground in the thinking of the early Wittgenstein, and is everywhere present in the writings of Heidegger, does indeed also play a role, albeit a less conspicuous one, in the work of the later Wittgenstein. Wittgenstein did not become unfaithful to the conclusion of the *Tractatus*. On the contrary, he adhered to it with great strictness.

[7] *Ibid.*, p. 376.

[8] C. F. Gethmann, "Philosophie als Vollzug und als Begriff. Heideggers Identitätsphilosophie des Lebens in der Vorlesung vom Wintersemester 1921/22 und ihr Verhältnis zu 'Sein und Zeit'", in: Frithjof Rodi (ed.), *Dilthey-Jahrbuch für Philosophie und Geschichte der Geisteswissenschaften* 4 (1986/87), p. 41.

[9] *Ibid.*, p. 42.

[10] Apel, *op. cit.*, p. 384. Apel does of course note that at the same time the conceptual edifice of *Being and Time* bears traces of that metaphysical realism whose avoidance should have been possible with the help of the concept of meaning Heidegger himself adheres to in that work. Heidegger's later thinking shows a clear recognition of this very state of affairs. Cf. Apel, *op. cit.*, p. 380.

[11] *Ibid.*, p. 394.

[12] In his paper "Heidegger and Ryle: Two Versions of Phenomenology", in: M. Murray (ed.), *Heidegger and Modern Philosophy*, New Haven: Yale University Press, 1978, p. 271–290.

[13] The review was, incidentally, reprinted in 1970–71 in the *Journal of the British Society for Phenomenology*.

[14] *Heidegger und Wittgenstein: Existential- und Sprachanalysen zu den Grundlagen philosophischer Anthropologie*, Stuttgart: Klett-Cotta, 1985, p. 228.

[15] *What Computers Can't Do: A Critique of Artificial Reason*, New York: Harper & Row, 1972, 2nd ed. 1979.

[16] Dreyfus, p. 233.

[17] *Ibid.*, p. 262.

[18] *Ibid.*, p. 57.

[19] *Antwort: Martin Heidegger im Gespräch*, ed. by Günther Neske and Emil Kettering, Pfullingen: Neske, 1988, p. 99.

[20] *Aus der Erfahrung des Denkens*, Frankfurt/M.: Vittorio Klostermann, 1983, pp. 134 f. The relevant aspects of *Being and Time* are set forth in detail in István Fehér,

"Heidegger und das Traditionsproblem", *Annales Universitatis Scientiarum Budapestinensis de Rolando Eötvös Nominatae – Sectio Philosophica et Sociologica XV* (1981), pp. 189–204. I am indebted to Prof. Fehér for his help in the preliminary research on the topic of this paper.

[21] From the Preface to *Philosophical Remarks*, Oxford: Basil Blackwell, 1975, p. 7.

[22] *Culture and Value*, p. 56e

[23] TS 213, p. 423, here quoted after the translation of von Wright in McGuinness, ed., *Wittgenstein and His Times*, Oxford: Blackwell, 1982, p. 113. The "idea of a way of thinking as reflecting the character of a culture" is, as McGuinness points out, part of the Sraffa heritage in Wittgenstein ("Freud and Wittgenstein", *ibid.*, p. 39).

[24] "Perhaps one day this civilization", wrote Wittgenstein in 1947, "will produce a culture. – When that happens there will be a real history of the discoveries of the 18th, 19th and 20th centuries, which will be deeply interesting", *Culture and Value*, p. 64.

[25] Written in 1947. *Vermischte Bemerkungen*, p. 120 (cf. *Culture and Value*, p. 63, translation amended).

[26] *Culture and Value*, pp. 48 f.

[27] "Hebel – der Hausfreund", *loc. cit.*, p. 146.

[28] Heidegger, *The Question Concerning Technology*, transl. William Lovitt, New York: Garland Publishing, 1977, p. 32.

[29] Cf. *Heraklit*, with Eugen Fink, Frankfurt/M.: Klostermann, 1970, p. 31.

[30] Personal communication by Brian McGuinness.

[31] Wittgenstein, *Schriften* 3, Frankfurt/M.: Suhrkamp, 1967, pp. 68 f., cf. also M. Murray's commentary in: Murray (ed.), *Heidegger and Modern Philosophy*, pp. 80 ff.

[32] And in one of the dictations to Schlick from 1931. Here the formula "das Nichts nichtet" is treated, not entirely without sympathy, as a metaphor capable of relieving a certain type of confusion (TS 302, p. 28).

[33] In this connection see also Allan Janik, "Why Is Wittgenstein Important?" in: R. Haller and J. Brandl, eds., *Wittgenstein – Towards a Re-Evaluation*, Vienna: Hölder-Pichler-Tempsky, 1991, vol. II.

[34] *Philosophical Investigations*, § 432.

[35] Ong (*Orality and Literacy*, p. 81) points to the "startling paradox" that writing has traditionally, ever since the time when the New Testament was composed, been closely associated with death: "The letter kills but the spirit gives life" (2 Cor 3,6). See also Ong, "Maranatha", in his *Interfaces of the Word: Studies in the Evolution of Consciousness and Culture*, Ithaca, N.Y.: Cornell University Press, 1977, pp. 233 ff., for a discussion of the wide range of texts that connect death and writing.

[36] "*ihm*, in der Mundart: 'eam', z.B.: 'I hob *eam* g'sogt' – *ihn*, in der Mundart: 'n' oder 'm', z.B.: 'I hob *m* g'sehn'", *Wörterbuch für Volksschulen* (1926 – repr. Vienna: Hölder-Pichler-Tempsky, 1977, eds. A. Hübner *et al.*), p. 15. – In the (originally unpublished) preface to the dictionary Wittgenstein makes the remark: "Again and again psychological principles (where will the student look for the word, how does one guard him against confusions in the best possible manner) clash with grammatical ones (base word, derivative) and with the typographical utilization of space, with the well-organized appearance of the printed page, etc.", *ibid.*, p. XXXV, transl. by Elisabeth Leinfellner.

[37] On this issue see, most importantly, the works of Eric Havelock, as well as the

study by Goody and Watt, "The Consquences of Literacy". Cf. also Walter J. Ong, SJ, *The Presence of the Word: Some Prolegomena for Cultural and Religious History*, New Haven: Yale University Press, 1967, and Ong, *Orality and Literacy*.

[38] "Wittgenstein", writes von Wright, "had done no systematic reading in the classics of philosophy. He could read only what he could wholeheartedly assimilate. ... as a young man he read Schopenhauer. From Spinoza, Hume, and Kant he said that he could only get occasional glimpses of understanding. I do not think that he could have enjoyed Aristotle or Leibniz, two great logicians before him. But it is significant that he did read and enjoy Plato." ("Ludwig Wittgenstein: A Biographical Sketch", in G. H. von Wright, *Wittgenstein*, Oxford: Basil Blackwell, 1982, p. 33.)

[39] "Die Bedeutung eines Wortes verstehen, heißt, einen Gebrauch kennen, verstehen", MS 111, p. 12.

[40] "Unsere Weise von den Wörtern zu reden können wir durch das beleuchten was Sokrates im 'Kratylos' sagt. Kratylos: 'Bei weitem & ohne Frage ist es vorzüglicher, Sokrates, durch ein ähnliches darzustellen, was jemand darstellen will, als durch das erste beste'. – Sokrates: 'Wohlgesprochen'."

[41] "Sokrates zu Theaitetos: 'Und wer vorstellt, sollte nicht etwas vorstellen?' – Th.: 'Notwendig'. – Soc.: 'Und wer etwas vorstellt, nichts Wirkliches?' – Th.: 'so scheint es'."

[42] "Wie undendlich einfach dieses Problem! Und wie seltsam, daß man es überhaupt als Problem konnte ansehen wollen", *ibid.*, p. 14, entry of July 14, 1931.

[43] "Ich finde bei Plato auf eine Frage wie 'was ist Erkenntnis' nicht die vorläufige Antwort: Sehen wir einmal nach, wie dieses Wort gebraucht wird", *ibid.*, pp. 26 f.

[44] "Was für großartige Menschen wir sind diese alten Probleme gelöst zu haben! – Nein die Zeit hat uns geändert & und die Probleme sind verschwunden", MS 154, p. 94.

[45] F. W. von Herrmann, *Die Selbstinterpretation Martin Heideggers*, Meisenheim am Glan: Verlag Anton Hain, 1964, p. 192.

[46] *What Is Called Thinking?*, Harper & Row, 1968, pp. 10 and 19.

[47] *Unterwegs zur Sprache*, Pfullingen: Neske, 1982, p. 266.

[48] *Ibid.*, p. 260.

[49] Martin Heidegger, *Being and Time*, transl. by John Macquarrie and Edward Robinson, Oxford: Basil Blackwell, 1962, p. 203.

[50] "Hearing is constitutive for discourse", *ibid.*, p. 206.

[51] *Ibid.*, pp. 205 and 204.

[52] "Communication is never anything like a conveying of experiences, such as opinions or wishes, from the interior of one subject into the interior of another", *ibid.*, p. 205.

[53] "When we are explicitly hearing the discourse of another, we proximally under- stand what is said, or – to put it more exactly – we are already with him, in advance, alongside the entity which the discourse is about. On the other hand, what we proximally hear is not what is voiced in the utterance. ... Both talking and hearing are based upon understanding" (*Being and Time*, pp. 207 f., translation slightly amended). Compare also, for instance, *What Is Called Thinking?*, p. 129: "When we hear directly what is spoken directly, we do not at first hear the words as terms, still less the terms as mere sound."

[54] As Heidegger remarks: "die Sprache ist nicht bloß Sprache, insofern wir diese ... als die Einheit von Lautgestalt (Schriftbild), Melodie und Rhythmus und Bedeutung (Sinn) vorstellen. Wir denken Lautgestalt und Schriftbild als den Wortleib, Melodie und Rhythmus als die Seele und das Bedeutungsmäßige als den Geist der Sprache" ("Brief über den 'Humanismus'", in: *Wegmarken*, Frankfurt/M.: Vittorio Klostermann, 1976, p. 333).

[55] "Logos", in: *Heidegger, Early Greek Thinking*, transl. by David Farrell Krell and Frank A. Capuzzi, San Francisco: Harper & Row, 1984, p. 77. It is significant that Heidegger does not question the widespread, but very probably false, assumption according to which the Presocratics were already philosophers of a literate culture (cf. e.g. the essay "Aletheia", *ibid.*, p. 102, where he speaks of "Heraclitus' writing"). On this problem see esp. Eric A. Havelock, "Preliteracy and the Presocratics", in: *The Literate Revolution in Greece and Its Cultural Consequences*.

[56] "Logos", *loc. cit.*, p. 66.

[57] "Hören", "Horchen", "Gehorchen", "Hörig". Cf. e.g. *Time and Being*, pp. 206 f.; *What Is Called Thinking?*, p. 48 (the mother will convey to her son "what obedience [das Gehorchen] is" by making him "hear" [hörend]); "Logos", *loc. cit.*, p. 65 ("hearing in the sense of hearkening and heeding ... a transposition of hearing proper into the realm of the spiritual [*das Geistige*]").

[58] "The Age of the World Picture", in: Heidegger, *The Question Concerning Technology and Other Essays*, transl. by William Lovitt, New York: Garland Publishing, 1977, p. 136. Heidegger then goes on to say: "The flight into tradition, out of a combination of humility and presumption, can bring about nothing in itself other than self-deception and blindness in relation to the historical moment."

[59] "The Question Concerning Technology", *ibid.*, p. 25.

[60] "daß der Mensch in einen Brauch gehöre, der ihn beansprucht", *Unterwegs zur Sprache*, pp. 125 f.

[61] *Early Greek Thinking*, p. 52.

[62] Cf. for instance *Time and Being*, p. 215; "Was ist Metaphysik", in: *Wegmarken*, p. 119; "Moira", in: *Early Greek Thinking*, p. 97 ("The essence of 'Αλήθεια remains veiled. The visibility it bestows allows the presencing of what is present to arise as outer appearance [*Aussehen*], ειδος and aspect [*Gesicht*], ιδέα. Consequently the perceptual relation to the presencing of what is present is defined as 'seeing' ειδέναι. Stamped with this character of visio [is] knowledge and the evidence of knowledge...").

[63] Cf. "The Age of the World Picture", *loc. cit.*, p. 131.

[64] Frankfurt: Vittorio Klostermann, 1983, pp. 68 f.

[65] Cf. for instance *Time and Being*, p. 140; "The Age of the World Picture", *loc. cit.*, p. 135; *Early Greek Thinking*, p. 17.

[66] Wittgenstein's infatuation with the cinema surely amounts to more than a simple biographical fact. His favourite distraction in Cambridge in the 1930s – the "flick", or, to spell it out, *the sound film* – was in fact an important audio-visual experience. Here the embeddedness of spoken language in extra-linguistic situations could hardly be overlooked. In Wittgenstein's remarks written in the early 1930s the film quite often plays the role of a metaphor or a simile. He notes, for example: "Das gesprochene Wort im Sprechfilm, das die Vorgänge auf der Leinwand begleitet, ist ebenso fliehend

// fliessend //, wie diese Vorgänge, und nicht das Gleiche wie der Tonstreifen" (TS 211, p. 708, cf. MS 113, p. 519).

[67] "We do not yet hear", writes Heidegger, "we whose hearing and seeing are perishing through radio and film under the rule of technology" (The *Question Concerning Technology*, p. 48).

[68] I refer here only to the book by the computer specialists T. Winograd and F. Flores, *Understanding Computers and Cognition: A New Foundation for Design*, Norwood, N.J.: Ablex, 1986. The number of philosophers interested in artificial intelligence and at the same time influenced by Heidegger is constantly increasing.

[69] "Are we the latecomers we are? But are we also at the same time precursors of the dawn of an altogether different age...?", *Early Greek Thinking*, p. 17.

NOTES TO CHAPTER 10

[1] London: Routledge & Kegan Paul, 1923.

[2] *Op. cit.*, p. 296.

[3] *Ibid.*, p. 306.

[4] *Ibid.*, pp. 307 and 311.

[5] *Ibid.*, pp. 312 and 307.

[6] *Ibid.*, p. 308. Ogden and Richards, too, associate Platonism with specific "linguistic habits" (*ibid.*, pp. 30 f.), without however recognizing the role writing here plays. The decisive breakthrough occurs in the works of Eric Havelock, a Cambridge graduate.

[7] I rather agree with Saul Kripke, according to whom the essentials of the private language argument precede §243. In §202, as Kripke writes, "the conclusion is already stated explicitly" (*Wittgenstein on Rules and Private Language*, Oxford: Blackwell, 1982, p. 3). And I think Kripke is right in his view that the argument is not primarily about "sensation language" (*ibid.*, p. 2).

[8] Rush Rhees, "Can There Be a Private Language?" (1954), repr. in G. Pitcher, ed., *Wittgenstein: The Philosophical Investigations*, New York: 1966, cf. esp. pp. 278 f.: "The point is that I speak a language that is spoken. What I say has significance in that language, not otherwise. ... For language there must be 'the way the expressions are used', and this goes with the way people live."

[9] G. P. Baker and P. M. S. Hacker, *Wittgenstein: Rules, Grammar and Necessity*. Oxford: Basil Blackwell, 1985, pp. 162, 164 and 179. This interpretation is in its turn criticized by Norman Malcolm in his "Wittgenstein on Language and Rules", *Philosophy* 64 (1989). "A rule", writes Malcolm, "can exist only in a human practice, or in what is analogous to it. And what a rule requires and what following it is, presupposes the background of a social setting in which there is quiet agreement as to what 'going on in the same way' is" (*loc. cit.*, p. 21). On this exchange and its implications see the valuable essay by Lars Hertzberg, "Wittgenstein and the Sharing of Language", in: R. Haller – J. Brandl, eds., *Wittgenstein – Towards a Re-Evaluation*, Vienna: 1990.

[10] With the exception of course of Kripke. In his interpretation, too, the problem of memory plays a decisive role; but he takes this problem to be a general one of rule-following.

[11] Compare also § 265: "Let us imagine a table (something like a dictionary) that

exists only in our imagination. A dictionary can be used to justify the translation of a word X into a word Y. But are we also to call it a justification if such a table is to be looked up only in the imagination? – 'Well, yes; then it is a subjective justification'. – But justification consists in appealing to something independent. – 'But surely I can appeal from one memory to another. For example, I don't know if I have remembered the time of departure of a train right, and to check it I call to mind how a page of the time-table looked. Isn't it the same here?' No; for this process has got to produce a memory which is actually *correct*. If the mental image of the time-table could not itself be *tested* for correctness, how could it confirm the correctness of the first memory? (As if someone were to buy several copies of the morning paper to assure himself that what it said was true.)

Looking up a table in the imagination is no more looking up a table than the image of the result of an imagined experiment is the result of an experiment".

[12] Or, really, to establish the identity or difference of single *letters*, to pass judgments of the sort "A = A". The correct *application* of "A" might depend on public criteria; the question of coherence *within* writing, whether A = A, is all the individual has to make decisions about.

[13] The stage is set in § 2 already, where the language-game described is a purely *spoken* one, consisting of orders, and indeed in § 1, where writing *does* play a role in the language-game introduced, but a role markedly subordinate to, weaved into the practice of, oral communication.

[14] Baker and Hacker, *op. cit.*, pp. 174 and 177.

[15] Kripke, *op. cit.*, pp. 8 f. and 19.

[16] Sellars, "Empiricism and the Philosophy of Mind", in: *Science, Perception and Reality*, p. 178.

[17] P. F. Strawson, "Review of Wittgenstein's *Philosophical Investigations*" (1954), in: G. Pitcher, ed., *Wittgenstein: The Philosophical Investigations. A Collection of Critical Essays*, Anchor Books, 1966, pp. 51 f. While accepting much of Strawson's argument, we do not adopt the terminology he here uses. The terms applied by Strawson – "inward speech", *ibid.*, p. 51, or "inner speech", *ibid.*, p. 52 – we reserve, as indicated above, for the discussion of a different dimension of language.

[18] As Strawson formulates it, "audible and inner speech are on the same level", *ibid.*, p. 52.

[19] *Philosophical Investigations*, Part II, pp. 220 f.

[20] Pitcher, ed., p. 52.

[21] *Ibid.*

[22] *Ibid.*, p. 44.

[23] *Thought and Language*, MIT, 1962, pp. 130 f.

[24] *Ibid.*, p. 131.

[25] *Ibid.*, p. 133.

[26] *Ibid.*, p. 134.

[27] *Ibid.*, p. 135.

[28] *Ibid.*, p. 135.

[29] *Ibid.*, p. 98.

[30] *Ibid.*, p. 144. This is the aspect stressed by Vygotsky's disciple Luria: "The writer must structure communication in a style that will enable the reader to be able to move

from the expanded, external speech to the inner sense of the text without relying on anything except the grammatical forms of the written communication" (Alexander R. Luria, *Language and Cognition*, Winston & Sons, 1981, p. 165).

[31] *Op. cit.*, p. 99.

[32] *Ibid.*, p. 146, cf. also pp. 147 f.

[33] "Formalized patterns of speech, recital under ritual conditions, the use of drums and other musical instruments, the employment of professional remembrancers – all such factors may shield at least part of the content of memory from the transmuting influence of the immediate pressures of the present". (Jack Goody and Ian Watt, "The Consequences of Literacy", 1963, in: Goody, ed., *Literacy in Traditional Societies*, Cambridge University Press, 1968, p. 31.)

[34] This might explain the apparent cognitive autonomy of autistic children, whose language indeed gives the impression, as the Hintikkas point out, of a private language in Wittgenstein's sense. (Merrill B. Hintikka – Jaakko Hintikka, "Wittgenstein über private Erfahrung", in: D. Birnbacher – A. Burkhardt, eds., *Sprachspiel und Methode: Zum Stand der Wittgenstein-Diskussion*, Berlin: de Gruyter, 1985, p. 10.) The language of autistic children consists, typically, in the "repetition of stored phrases", coupled, at a very early age already, with "unusual facility in rote memory", "the ability to repeat endless numbers of rhymes, catechisms, lists of names" (L. Eisenberg – L. Kanner, "Early Infantile Autism, 1943–1955", *American Journal of Orthopsychiatry* 26, 1956, pp. 556 f.). In other respects, too, they are obsessed by repetition and sameness – "once having accepted a new pattern" they would "incorporate it into the restricted set of rituals which then had to be endlessly iterated" (*ibid.*, p. 557). Still, autistic children certainly do not create the impression of being devoid of thought processes; the impression their parents gain is, rather, that of "a mental barrier between ... inner consciousness and the outside world" (L. Eisenberg, "The Autistic Child in Adolescence", *American Journal of Psychiatry* 112, 1956, p. 609). The language of the autistic child, writes Leo Kanner, possesses its "own private, original, individualized" meanings, a "private, original frame of reference" (Kanner, "Irrelevant and Metaphorical Language in Early Infantile Autism", *American Journal of Psychiatry* 103, 1946, pp. 243 f.). These meanings turn out to rest on metaphor, analogical transfer, the sources of which are "rooted in *concrete, specific, personal* experiences of the child". The idiosyncratic expressions are intelligible "only to the extent to which any listener can, through his own efforts, trace the source of the analogy", i.e. only to a severely limited extent (*ibid.*, p. 243). Nor are these expressions applied with "the purpose of communication" (Eisenberger – Kanner, p. 556). Autistic thinking, one might say, is anchored in excessive, self-contained oral repetition rather than in collective speech habits. Writing/reading, too, can serve as a substitute for public speech; significantly, autistic children often display good reading capacities (see e.g. Kanner, "Irrelevant and Metaphorical Language", p. 243, compare also Bruno Bettelheim, *The Empty Fortress: Infantile Autism and the Birth of the Self*, New York: The Free Press, 1967, pp. 219 f. Note that the view I am here putting forward retains its force even if autism is seen as the result of "preexisting neurologic dysfunction" rather than of exclusively "environmental causes", cf. Bemporad *et al.*, "Autism and Emotion: An Ethological Theory", *American Journal of Orthopsychiatry* 57/4, 1987, p. 481).

[35] Cf. Walter J. Ong, *The Presence of the Word: Some Prolegomena for Cultural and Religious History*, New Haven: Yale University Press, 1967, pp. 117, 127, 149.

[36] Maurice Halbwachs' *Les Cadres sociaux de la mémoire* (1925) is obviously relevant here. But note that Halbwachs did *not* explicitly distinguish, as Goody and Watt, building on his insights, later indeed did, between oral and literate societies. In non-literate societies, write Goody and Watt, "language is developed in intimate association with the experience of the community, and is learned by the individual in face-to-face contact with the other members. What continues to be of social relevance is stored in the memory while the rest is usually forgotten" (Goody – Watt, *loc. cit.*, pp. 30 f.).

[37] Merrill B. Hintikka – Jaakko Hintikka, *Investigating Wittgenstein*, Oxford: Blackwell, 1986, p. 275.

[38] See esp. chapter 2 of the present volume, as well as my papers "Wittgenstein's Later Work in relation to Conservatism", in: Brian McGuinness, ed., *Wittgenstein and his Times*, Oxford: Blackwell, 1982, pp. 44–68, and "Musil und Wittgenstein", *Literatur und Kritik* 113 (1977/4), repr. in Nyíri, *Gefühl und Gefüge*.

[39] As Goody and Watt put it, "writing, by objectifying words, and by making them and their meaning available for much more prolonged and intensive scrutiny than is possible orally, encourages private thought; the diary or the confession enables the individual to objectify his own experience, and gives him some check upon the transmutations of memory under the influences of subsequent events" (*loc. cit.*, p. 62).

[40] Cf. e.g. his contribution in O. R. Jones, ed., *The Private Language Argument*, London: 1971, pp. 136 ff.

[41] For example, my own understanding of the connotations of "Gregor Samsa" is obviously influenced by the fact that I have read Kafka's fragment "Hochzeitsvorbereitungen auf dem Lande", as well as his "Brief an den Vater", practically at the same time as "Die Verwandlung", In the fantasies of the protagonist of "Hochzeitsvorbereitungen", his seeking seclusion is associated with the idea of his becoming a cockchafer or a stag-beetle; in the letter Kafka writes of his habit of hiding away from his father – the term, *sich verkriechen*, means also "crawl away", "creep into a hole". I can explain to another, and thereby perhaps have him share, my particular way through Kafka; but not, clearly, all those literary and linguistic experiences of mine that prepared, and surround, that way.

[42] What Jerome S. Bruner says in his "Introduction" to *Thought and Language* is relevant here: "Vygotsky has ... introduced an historical perspective into the understanding of how thought develops, and indeed what thought is. But what is interesting is that he has also proposed a mechanism whereby one becomes free of one's history. ... To me, the striking fact is that given a pluralistic world where each comes to terms with the environment in his own style, Vygotsky's developmental theory is also a description of the many roads to individuality and freedom", *op. cit.*, pp. ix f.

BIBLIOGRAPHY

Acton, H. B., "Tradition and Some Other Forms of Order", *Proceedings of the Aristotelian Society*, N.S., vol. LIII (1953).

Alloni-Fainberg, Y., "Official Hebrew Terms for Parts of the Car", in Joshua A. Fishman, ed., *Advances in Language Planning*, The Hague: Mouton, 1974.

Allport, Gordon W., *The Nature of Prejudice*, Garden City, N.Y.: Doubleday, 1958.

Ambrose, Alice, and Morris Lazerowitz (eds.), *Ludwig Wittgenstein: Philosophy and Language*, London: George Allen and Unwin, 1972.

Apel, Karl-Otto, "Wittgenstein und Heidegger. Die Frage nach dem Sinn von Sein und der Sinnlosigkeitsverdacht gegen alle Metaphysik" (1967), in: O. Pöggeler (ed.), *Heidegger: Perspektiven zur Deutung seines Werks*, Königstein/Ts.: Athenäum, 1984.

Armstrong, D. M., "The Nature of Tradition", in Armstrong, *The Nature of Mind*, Harvester Press, 1981.

Baeck, Leo, "Die jüdische Religion in der Gegenwart", *Süddeutsche Monatshefte* 27 (Sept. 1930).

Baeck, Leo, *Das Wesen des Judentums*, 6th ed., Köln: Melzer, 1960.

Baker, G. P., and P. M. S. Hacker, *Wittgenstein: Rules, Grammar and Necessity*. Oxford: Basil Blackwell, 1985.

Baldamus, W., "Ludwig Fleck and the Development of the Sociology of Science", in: Peter R. Gleichmann, Johan Goudsblom, and Hermann Korte (eds.), *Human Figurations: Essays for Norbert Elias*, Amsterdam: Stichting Amsterdams Sociologisch Tijdschrift, 1977, pp. 135–156.

Bárczi, Géza, *A magyar nyelv életrajza*, Budapest: Gondolat, 1975.

Bárczi, Géza, "Az eszperantó nyelvi értéke", *Eszperantó Magazin* (Budapest: 1975).

Bell, Daniel, "The Social Framework of the Information Society", in M. L. Dertouzos and Joel Moses, eds., *The Computer Age: A Twenty-Year View*, Cambridge, Mass.: MIT Press, 1979.

Bemporad *et al.*, "Autism and Emotion: An Ethological Theory", *American Journal of Orthopsychiatry* 57/4, 1987.

Benedict, Ruth, *Patterns of Culture*, Boston: 1934.

Bennett, Neville, *Teaching Styles and Pupil Progress*, London: Open Books, 1976.

Bettelheim, Bruno, *The Empty Fortress: Infantile Autism and the Birth of the Self*, New York: The Free Press, 1967.

Bloomfield, Leonard, *Language*, New York: 1933.

Bloomfield, Leonard, "Meaning", *Monatshefte für Deutschen Unterricht*, vol. 35 (1943).

Bloor, D., "Wittgenstein and Mannheim on the Sociology of Mathematics", *Studies in the History and Philosophy of Science*, vol. 4 (1973).

Bloor, D., "The Strengths of the Strong Programme" (1981), in: J.R.Brown, ed.,

Scientific Rationality: The Sociological Turn, Dordrecht: Reidel, 1984.

Bolinger, Dwight, "The Atomization of Meaning", *Language*, vol. 41 (1965).

Bolinger, Dwight, "Meaning and Memory", in George G. Haydu, ed., *Experience Forms: Their Cultural and Individual Place and Function*, The Hague: Mouton, 1979.

Bolkosky, S. M., *The Distorted Image: German Jewish Perceptions of Germans and Germany, 1918–1935*, New York: Elsevier, 1975.

Bollnow, O. Fr., "Wilhelm von Humboldts Sprachphilosophie", *Zeitschrift für deutsche Bildung*, vol. 15 (1938).

Bolter, David, *Turing's Man: Western Culture in the Computer Age*, Chapel Hill: The University of North Carolina Press, 1984.

Brugmann, K., and August Leskien, *Zur Kritik der künstlichen Weltsprachen*, Straßburg: 1907.

Buber, Martin, *Der Jude und sein Judentum: Gesammelte Aufsätze und Reden*, Köln: Melzer, 1963.

Burke, Edmund, "A Philosophical Inquiry into the Origin of our Ideas of the Sublime and Beautiful" (1756/57), in: Burke, *The Works: Twelwe Volumes in Six*, 1887, vol. I.

Burke, Edmund, "Reflections on the Revolution in France" (1790), in: Burke, *The Works: Twelwe Volumes in Six*, 1887, Vol. III.

Cahnman, Werner J., "Sociology", in: *Encyclopaedia Judaica*, Jerusalem: n.d.

Carnap, Rudolf, "Erwiderung auf die vorstehenden Aufsätze von E. Zilsel und K. Duncker". *Erkenntnis* 3 (1932/33).

Chafe, Wallace L., "Idiomaticity as an Anomaly in the Chomskyan Paradigm", *Foundations of Language*, vol. 4 (1968).

Clark, Katerina and Michael Holquist, *Mikhail Bakhtin*, Cambridge, Mass.: Harvard University Press, 1984.

Cohen, Robert S. and Thomas Schnelle (eds.), *Cognition and Fact: Materials on Ludwik Fleck*, Dordrecht: Reidel, 1986.

Coleman, Samuel, "Is There Reason in Tradition?", in: Preston King and B.C. Parekh, eds., *Politics and Experience*, Cambridge: Cambridge University Press, 1968.

Coser, Lewis A., "Georg Simmel's Style of Work: A Contribution to the Sociology of the Sociologist", *The American Journal of Sociology* 63 (1957–1958), pp. 635–641.

Deneffe, August, SJ, *Der Traditionsbegriff*, Münster/Westf.: 1931.

Deutsch, Karl W., *Nationalism and Social Communication: An Inquiry into the Foundations of Nationality*, London: 1953.

Dewey, John, *Human Nature and Conduct*, New York: The Modern Library, 1922.

Dijkstra, Edsger W., *Selected Writings on Computing: A Personal Perspective*, New York: Springer, 1982.

Dimler, G. Richard, S. J., "Word Processing and the New Electronic Language", *Thought*, vol. 61, no.243 (Dec. 1986).

Dostojewsky, F. M., *Die Dämonen*, R. Piper, 1921.

Douglas, Mary, *How Institutions Think*, Syracuse, N.Y.: Syracuse University Press, 1986.

Dretske, Fred I., *Knowledge and the Flow of Information*, Oxford: Basil Blackwell,

1981.

Dreyfus, Hubert L., *What Computers Can't Do: A Critique of Artificial Reason*, New York: Harper & Row, 1972, 2nd ed. 1979.

Durkheim, Emile, *The Division of Labor in Society* (1893), transl. by G. Simpson, New York: The Free Press, 1964.

Durkheim, Émile, *Le Suicide: Étude de Sociologie*, Paris: 1897.

Durkheim, Émile, *Les formes élémentaires de la vie religieuse*, Paris: 1912.

Durkheim, Émile, *Pragmatisme et sociologie*, Paris: 1981.

Eisenberg, L., and L. Kanner, "Early Infantile Autism, 1943–1955", *American Journal of Orthopsychiatry* 26, 1956.

Eisenberg, L., "The Autistic Child in Adolescence", *American Journal of Psychiatry* 112, 1956.

Eisenstadt, S. N., *Tradition, Change, and Modernity*, New York: Wiley, 1973.

Eisenstein, Elizabeth, *The Printing Press as an Agent of Change: Communications and Cultural Transformations in Early-Modern Europe*, Cambridge: Cambridge University Press, 1979.

Eliot, T. S., *The Sacred Wood: Essays on Poetry and Criticism* (1920), London: Methuen, 1960.

Engelmann, Paul, *Letters from Ludwig Wittgenstein: With a Memoir*, Oxford: Basil Blackwell, 1967.

Epstein, Klaus, *The Genesis of German Conservatism*, Princeton, N.J.: Princeton University Press, 1966.

Fann, K.T., ed., *Wittgenstein: The Man and His Philosophy*, New York: Dell, 1967.

Fehér, István, "Heidegger und das Traditionsproblem". *Annales Universitatis Scientiarum Budapestinensis de Rolando Eötvös Nominatae – Sectio Philosophica et Sociologica* XV (1981).

Feigenbaum, Edward A., and Pamela McCorduck, *The Fifth Generation*, New York: Signet, 1984.

Ferguson, C.A., "Sociolinguistic Settings of Language Planning", in Joan Rubin *et al.*, eds., *Language Planning Processes*, The Hague: Mouton, 1977.

Feyerabend, Paul, *Science in a Free Society* (1978). London: Verso, 1982.

Finley, M. I., "Myth, Memory, and History", *History and Theory* 1965.

Fleck, Ludwik, *Genesis and Development of a Scientific Fact*, Chicago: University of Chicago Press, 1979.

Gadamer, Hans-Georg, *Truth and Method*, New York: Crossroad, 1975.

Garver, Newton, "Die Lebensform in Wittgensteins Philosophischen Untersuchungen", *Grazer Philosophische Studien*, vol. 21 (1984).

Gehlen, Arnold, *Der Mensch*, Berlin: Junker und Dünnhaupt, 1940.

Gellner, Ernest, "Nationalism", in Gellner, *Thought and Change*, London: Weidenfeld & Nicolson, 1964.

Gethmann, C. F., "Philosophie als Vollzug und als Begriff. Heideggers Identitätsphilosophie des Lebens in der Vorlesung vom Wintersemester 1921/22 und ihr Verhältnis zu 'Sein und Zeit'", in: Frithjof Rodi (ed.), *Dilthey-Jahrbuch für Philosophie und Geschichte der Geisteswissenschaften* 4 (1986/87).

Goodstein, R. L., "Wittgenstein's Philosophy of Mathematics", in: Alice Ambrose and Morris Lazerowitz (eds.), *Ludwig Wittgenstein: Philosophy and Language*,

London: George Allen and Unwin, 1972.

Goody, Jack and Ian Watt, "The Consequences of Literacy", 1963, in: Goody (ed.), *Literacy in Traditional Societies*, Cambridge: Cambridge University Press, 1968.

Greenberg, Joseph H., *Essays in Linguistics*, New York: Wenner-Gren, 1957.

Grimm, J., and W. Grimm, *Deutsches Wörterbuch*, Leipzig: 1852.

Groningen, B.A. van, *In the Grip of the Past: Essay on an Aspect of Greek Thought*, Leiden: E.J. Brill, 1953.

Gumplowicz, Ludwig, *Der Rassenkampf: Sociologische Untersuchungen*, Innsbruck: 1883.

Gumplowicz, Ludwig, *Grundriß der Sociologie*, Wien: 1885.

Gumplowicz, Ludwig, *Soziologische Essays*, Innsbruck: 1899.

Haefner, Klaus, *Die neue Bildungskrise: Lernen im Computerzeitalter*, Reinbek bei Hamburg: Rowohlt, 1985.

Hahn, F., "Zur Verschriftlichung mündlicher Tradition in der Bibel", *Zeitschrift für Religions- und Geistesgeschichte* 39, 1987.

Hajnal, István, "Le rôle social de l'écriture et l'évolution européenne", *Revue de l'Institut de Sociologie* (Bruxelles), 1934.

Hajnal, István, "A technika fejlödése", in: *Domanovszky Emlékkönyv*, 1937.

Hajnal, István, "Universities and the Development of Writing in the XIIth–XIIIth Centuries", *Scriptorum. International Review of Manuscript Studies*, VI/2, 1952.

Halbwachs, Maurice, *Les cadres sociaux de la mémoire* (1925), new ed. Paris: 1975.

Halbwachs, Maurice, *La mémoire collective*, 1950, 2nd ed., Paris: 1968.

Hall, R.A., *Linguistics and Your Language*, New York: Doubleday, 1960.

Haller, Rudolf, "Über das sogenannte Münchhausentrilemma", *Ratio* 16 (1974).

Haller, Rudolf (ed.), *Schlick und Neurath – Ein Symposion*, Amsterdam: Rodopi, 1982 (*Grazer Philosophische Studien* 16/17).

Haller, Rudolf, *Urteile und Ereignisse: Studien zur philosophischen Logik und Erkenntnistheorie*, Freiburg: Verlag Karl Alber, 1982.

Haller, Rudolf, "Lebensform oder Lebensformen?", *Grazer Philosophische Studien*, vol. 21 (1984).

Haller, Rudolf, *Fragen zu Wittgenstein und Aufsätze zur Österreichischen Philosophie*, Amsterdam: Rodopi, 1986.

Hart, H. L. A., *The Concept of Law*, 2nd ed., Oxford: Clarendon Press, 1963.

Haugen, Einar, "Instrumentalism in Language Planning", in Joan Rubin and Björn H. Jernudd, eds., *Can Language Be Planned?*, The University Press of Hawaii, 1971.

Haugen, Einar, "Linguistics and Language Planning", in E. Haugen, *Studies*, ed. by E. Sch. Firchow *et al.*, The Hague: Mouton, 1972.

Hauser, Arnold, *The Sociology of Art*, transl. by Kenneth J. Northcott, London: Routledge & Kegan Paul, 1974.

Havelock, Eric, *Preface to Plato*, Cambridge, Mass.: Harvard University Press, 1963.

Havelock, Eric, *The Greek Concept of Justice: From Its Shadow in Homer to Its Substance in Plato*, Cambridge, Mass.: Harvard University Press, 1978.

Havelock, Eric, *The Literate Revolution in Greece and Its Cultural Consequences*, Princeton University Press, 1982.

Havelock, Eric, *The Muse Learns to Write: Reflections on Orality and Literacy from Antiquity to the Present*, New Haven: Yale University Press, 1986.

Hayek, F. A. von, "Individualism: True and False", in: Hayek, *Individualism and Economic Order* (1949), London: Routledge & Kegan Paul, 1976.

Hayek, F. A. von, *Studies in Philosophy, Politics and Economics*, London: Routledge & Kegan Paul, 1967.

Hayek, F. A. von, *Law, Legislation and Liberty*, vol. 3, London: Routledge & Kegan Paul, 1979.

Hebel, J. P., *Werke*, Karlsruhe: 1847.

Heidegger, Martin, *Being and Time*, transl. by John Macquarrie and Edward Robinson, Oxford: Basil Blackwell, 1962.

Heidegger, Martin, *What Is Called Thinking?*, transl. by J. Glenn Gray, Harper & Row, 1968.

Heidegger, Martin, *Heraklit*, with Eugen Fink, Frankfurt/M.: Klostermann, 1970.

Heidegger, Martin, "Brief über den 'Humanismus'", in: *Wegmarken*, Frankfurt/M.: Vittorio Klostermann, 1976.

Heidegger, Martin, *The Question Concerning Technology*, transl. by William Lovitt, New York: Garland Publishing, 1977.

Heidegger, Martin, *Unterwegs zur Sprache*, Pfullingen: Neske, 1982.

Heidegger, Martin, *Aus der Erfahrung des Denkens*, Frankfurt/M.: Klostermann, 1983.

Heidegger, Martin, *Einführung in die Metaphysik*, Frankfurt: Vittorio Klostermann, 1983.

Heidegger, Martin, *Early Greek Thinking*, transl. by David Farrell Krell and Frank A. Capuzzi, San Francisco: Harper & Row, 1984.

Heidegger, Martin, *Antwort: Martin Heidegger im Gespräch*, ed. by Günther Neske and Emil Kettering, Pfullingen: Neske, 1988.

Hempel, Carl G., "On the Logical Positivists' Theory of Truth", *Analysis* 2 (1935), pp. 49–59.

Herrmann, F. W. von, *Die Selbstinterpretation Martin Heideggers*, Meisenheim am Glan: Verlag Anton Hain, 1964.

Hertzberg, Lars, "Wittgenstein and the Sharing of Language", in: R. Haller – J. Brandl, eds., *Wittgenstein – Towards a Re-Evaluation*, Vienna: 1990.

Heyse, K. W. L., *System der Sprachwissenschaft*, Berlin: 1856.

Hintikka, Merrill B., and Jaakko Hintikka, "Wittgenstein über private Erfahrung", in: D. Birnbacher – A. Burkhardt, eds., *Sprachspiel und Methode: Zum Stand der Wittgenstein-Diskussion*, Berlin: de Gruyter, 1985.

Hintikka, Merrill B., and Jaakko Hintikka, *Investigating Wittgenstein*, Oxford: Blackwell, 1986.

Hobsbawm, Eric, and Terence Ranger, eds., *The Invention of Tradition*, Cambridge: Cambridge Univ. Press, 1983.

Hockett, Charles F., *The State of the Art*, The Hague: Mouton, 1968.

Holenstein, Elmar, *Menschliches Selbstverständnis*, Frankfurt/M.: Suhrkamp, 1985.

Hollinger, David, "T.S. Kuhn's Theory of Science and Its Implications for History", in: Gary Gutting, ed., *Paradigms and Revolutions: Appraisals and Applications of Thomas Kuhn's Philosophy of Science*, Notre Dame: University of Notre Dame Press, 1980.

Innis, Harold A., *The Bias of Commmunication*, University of Toronto Press, 1951, repr. 1971.

Janik, Allan, "Why Is Wittgenstein Important?", in: R. Haller and J. Brandl, eds., *Wittgenstein – Towards a Re-Evaluation*, Vienna: Hölder-Pichler-Tempsky, 1991, vol. II.

Joas, Hans, "Durkheim und der Pragmatismus. Bewußtseinspsychologie und die soziale Konstitution der Kategorien", in: Émile Durkheim, *Schriften zur Soziologie der Erkenntnis*, Frankfurt/M.: Suhrkamp, 1987.

Johnson, Samuel, *Dictionary of the English Language*, 1755.

Johnston, William M., *The Austrian Mind: An Intellectual and Social History, 1848–1938*, Berkeley: University of California Press, 1972.

Jones, O. R., ed., *The Private Language Argument*, London: 1971.

Jünger, Ernst, "Über Nationalismus und Judenfrage", *Süddeutsche Monatshefte* 27 (Sept. 1930).

Kaltenbrunner, Gerd-Klaus, "Der schwierige Konservatismus", in: Kaltenbrunner, ed., *Rekonstruktion des Konservatismus*, Freiburg i.B.: Rombach, 1972.

Kanner, Leo, "Irrelevant and Metaphorical Language in Early Infantile Autism", *American Journal of Psychiatry* 103, 1946.

Kant, Immanuel, *Critique of Pure Reason*, transl. by N. K. Smith, 2nd ed., London: 1933.

Katz, J. J., and J.A. Fodor, "The Structure of a Semantic Theory", *Language*, vol. 39 (1963).

Kaulla, Rudolf, *Der Liberalismus und die deutschen Juden: Das Judentum als konservatives Element*, Leipzig: Duncker und Humblot, 1928.

Kenny, Anthony, *Wittgenstein*, Harmondsworth, Middlesex: Penguin Books, 1973.

King, M. D., "Reason, Tradition, and the Progressiveness of Science" (1971), repr. in Gary Gutting, ed., *Paradigms and Revolutions: Appraisals and Applications of Thomas Kuhn's Philosophy of Science*, Notre Dame: University of Notre Dame Press, 1980.

Klemperer, Klemens von, *Germany's New Conservatism: Its History and Dilemma in the Twentieth Century*, Princeton, N.J.: Princeton University Press, 1957.

Koselleck, Reinhart, *Vergangene Zukunft: Zur Semantik geschichtlicher Zeiten*, Frankfurt/M.: Suhrkamp, 1979.

König, René, "Émile Durkheim", in: Dirk Käsler (ed.), *Klassiker des soziologischen Denkens*, vol. 1, München: C. H. Beck, 1976, pp. 312–364.

Kripke, Saul, *Wittgenstein on Rules and Private Language*, Oxford: Blackwell, 1982.

Kuhn, Thomas S., *The Structure of Scientific Revolutions*, The University of Chicago Press, 1962.

Kuhn, Thomas S., "Reflections on my Critics", in: Imre Lakatos and Alan Musgrave (eds.), *Criticism and the Growth of Knowledge*, Cambridge: Cambridge University Press, 1970.

Kuhn, Thomas S., *The Essential Tension: Selected Studies in Scientific Tradition and Change*, Chicago: The University of Chicago Press, 1977.

Kümmel, W.G., *Heilsgeschehen und Geschichte*, Marburg: 1965.

Leiser, Burton M., *Custom, Law, and Morality: Conflict and Continuity in Social Behavior*, Garden City, N.Y.: Anchor Books, 1969.

Leoni, Bruno, *Freedom and the Law*, Princeton, N.J.: Van Nostrand, 1961.
Levi, A. W., "Wittgenstein as Dialectician", in: K. T. Fann, ed., *Wittgenstein: The Man and His Philosophy*, New York: Dell, 1967.
Levy, David, "Constraining the Choice Set: Lessons from the Software Revolution", *Reason Papers*, Spring 1985.
Lewis, David K., *Convention: A Philosophical Study*, Cambr., Mass.: Harvard Univ. Press, 1969.
Lukács, Georg, *Dostojewski: Notizen und Entwürfe*, ed. J. C. Nyíri, Budapest: Akadémiai, 1985.
Lukács, Georg, *Geschichte und Klassenbewußtsein* (1923), Neuwied: Luchterhand, 1968.
Lukes, Steven, *Émile Durkheim: His Life and Work*, London: Allen Lane, 1973.
Luria, Alexander R., *Language and Cognition*, Winston & Sons, 1981.
Lübbe, Hermann, "Rationalitätsverluste. Über Tendenzen der Flucht aus der Gegenwart", *Jahrbuch 1982 der Technischen Universität München*.
Lübbe, Hermann, "'Neo-Konservative' in der Kritik", *Merkur* 1983/6.
Lübbe, Hermann, *Zeit-Verhältnisse: Zur Kulturphilosophie des Fortschritts*, Graz – Wien – Köln: Styria, 1983.
Lyotard, Jean-François, *La condition postmoderne*, Paris: 1979.
Mach, Ernst, *Popular Scientific Lectures*, transl. by Thomas J. McCormack, La Salle, Ill.: Open Court, 1986.
McCagg, William O., Jr., *Jewish Nobles and Geniuses in Modern Hungary*, Boulder: East European Quarterly, 1972.
McGuinness, Brian, "Freud and Wittgenstein", in McGuinness, ed., *Wittgenstein and His Times*, Oxford: Blackwell, 1982.
McLellan, David, *Karl Marx: His Life and Thought*, London: Macmillan, 1973.
McLuhan, Marshall, *Understanding Media: The Extensions of Man*, New York: McGraw-Hill, 1964.
Malcolm, Norman, "Wittgenstein on Language and Rules", *Philosophy* 64 (1989).
Malinowski, Bronislaw, "The Problem of Meaning in Primitive Languages", in C. K. Ogden and I. A. Richards, *The Meaning of Meaning: A Study of the Influence of Language upon Thought and of the Science of Symbolism*, London: Routledge & Kegan Paul, 1923.
Mann, Thomas, "Russische Anthologie", in: Mann, *Rede und Antwort: Gesammelte Abhandlungen und kleine Aufsätze*, Berlin: S. Fischer, 1925.
Mannheim, Karl, "Das konservative Denken", *Archiv für Sozialwissenschaft und Sozialpolitik*, vol. 57 (1927).
Marx, Karl, "Das philosophische Manifest der historischen Rechtsschule", 1842. Marx–Engels, *Werke*, vol. 1, Berlin: Dietz, 1964.
Marx, Karl, *The Poverty of Philosophy*, Moscow: n.d.
Marx, Karl, *Zur Kritik der politischen Ökonomie*. Marx–Engels, *Werke*, vol. 13, Berlin: Dietz, 1971.
Marx, Karl, *Foundations of the Critique of Political Economy*, transl. by Martin Nicolaus, Penguin Books, 1973.
Marx, Karl, *Das Kapital*, vol. 1, Berlin: Dietz, 1969.
Marx, Karl, *Das Kapital*, vol. 3, Berlin: Dietz, 1969.

Menger, Carl, *Untersuchungen über die Methode der Socialwissenschaften* (1883), in *The Collected Works of Carl Menger*, ed. F. A. von Hayek, vol. 2, London: The London School of Economics, 1934.

Merkel, Reinhard, "Das Klappen der Schere des Haarschneiders", *Die Zeit*, Sept. 1, 1989.

Merton, Robert K., "The Sociology of Knowledge", in: Georges Gurvitch and Wilbert E. Moore (eds.), *Twentieth Century Sociology*, New York: The Philosophical Library, 1945, pp. 366–405.

Moss, Gordon E., "Identification and the Curve of Optimal Cohesion", in: George G. Haydu, *Experience Forms: Their Cultural and Individual Place and Function*, The Hague: Mouton, 1979.

Murray, Michael, "Heidegger and Ryle: Two Versions of Phenomenology", in: Murray (ed.), *Heidegger and Modern Philosophy*, New Haven: Yale University Press, 1978.

Musil, Robert, *The Man Without Qualities* (1930), transl. by E. Wilkins and E. Kaiser, London: Panther Books, 1968.

Neurath, Otto, *Empiricism and Sociology*, ed. M. Neurath and R.S. Cohen, Dordrecht: Reidel, 1973.

Neurath, Otto, *Gesammelte philosophische und methodologische Schriften*, ed. R. Haller and H. Rutte, Wien: Hölder-Pichler-Tempsky, 1981.

Németh, László, *Kisebbségben*, Kecskemét: 1939.

Nietzsche, Friedrich, "Vom Nutzen und Nachteil der Historie für das Leben", *Unzeitgemäße Betrachtungen. Sämtliche Werke: Kritische Studienausgabe*, vol. 1.

Nyíri, J. C., "Wittgenstein's Later Work in Relation to Conservatism", in: Brian McGuinness, ed., *Wittgenstein and His Times*, Oxford: Basil Blackwell, 1982.

Nyíri, J. C., *Gefühl und Gefüge: Studien zum Entstehen der Philosophie Wittgensteins*, Amsterdam: Rodopi, 1986.

Nyíri, J. C., *Am Rande Europas: Studien zur österreichisch-ungarischen Philosophiegeschichte*, Wien: Böhlau, 1988.

Nyíri, J. C., "Wittgenstein and the Problem of Machine Consciousness", *Grazer Philosophische Studien* 33/34 (1989).

Oakeshott, Michael, *Rationalism in Politics*. London: Methuen, 1962.

Ong, Walter J., SJ, *The Presence of the Word: Some Prolegomena for Cultural and Religious History*. New Haven: Yale University Press, 1967.

Ong, Walter J., SJ, *Orality and Literacy: The Technologizing of the Word*, London: Methuen, 1982.

Park, Robert E., "Reflections on Communication and Culture", *The American Journal of Sociology*, Sept. 1938.

Pascal, Fania, "Wittgenstein: A Personal Memoir", *Encounter*, vol. 16, August 1973.

Pieper, Josef, "Über den Begriff der Tradition", *Arbeitsgemeinschaft für Forschung des Landes Nordrhein-Westfalen: Geisteswissenschften*, Heft 72, Köln: Westdeutscher Verlag, 1958.

Plumb, J. H., *The Death of the Past*, London: Macmillan, 1969.

Pocock, J. G. A., "Time, Institutions and Action: An Essay on Traditions and Their Understanding", in: Preston King and B.C. Parekh, eds., *Politics and Experience*, Cambridge: Cambridge University Press, 1968.

Polanyi, Michael, *Science, Faith and Society* (1946). Enl. ed., Chicago: University of Chicago Press, 1964.

Polanyi, Michael, *Personal Knowledge: Towards a Post-Critical Philosophy* (1958). London: Routledge & Kegan Paul, 1983.

Popper, Karl R., *Conjectures and Refutations*, London: Routledge and Kegan Paul, 1972.

Popper, Karl R., *Unended Quest: An Intellectual Autobiography*, rev. ed., Fontana/Collins, 1976.

Radin, Max, "Tradition", in *Encyclopaedia of the Social Sciences*, New York: Macmillan, 1935, vol. 15.

Renan, E., *Geschichte des Volkes Israel*, transl. E. Schaelsky, Berlin: Cronbach, 1894.

Rentsch, Thomas, *Heidegger und Wittgenstein: Existential- und Sprachanalysen zu den Grundlagen philosophischer Anthropologie*, Stuttgart: Klett-Cotta, 1985.

Rhees, Rush, "Can There Be a Private Language?" (1954), repr. in G. Pitcher, ed., *Wittgenstein: The Philosophical Investigations*, New York: 1966.

Rösler, Wolfgang, "Schriftkultur und Fiktionalität. Zum Funktionswandel der griechischen Literatur von Homer bis Aristoteles", in: A. u. J. Assmann – Chr. Hardmeier, eds., *Schrift und Gedächtnis: Archäologie der literarischen Kommunikation I*, München: 1983.

Ryle, Gilbert, *The Concept of Mind*, London: Hutchinson's University Library, 1949.

Sampson, Geoffrey, *Liberty and Language*, Oxford: Oxford University Press, 1979.

Savigny, Eike von, "Viele gemeinsame menschliche Handlungsweisen", in: W.L. Gombocz *et al.*, eds., *Traditionen und Perspektiven der analytischen Philosophie: Festschrift für Rudolf Haller*, Wien: Hölder-Pichler-Tempsky, 1989.

Schaub, Edward L., "A Sociological Theory of Knowledge", *The Philosophical Review* 29 (1920), pp. 319–339.

Scheler, Max (ed.), *Versuche zu einer Soziologie des Wissens*, München: Duncker & Humblot, 1924.

Scheler, Max, *Formalism in Ethics and Non-Formal Ethics of Values*, transl. M. S. Frings and R. L. Funk, Evanston: Northwestern University Press, 1973.

Schwemmer, Oswald, "Aspekte der Handlungsrationalität. Überlegungen zur historischen und dialogischen Struktur unseres Handelns", in: H. Schnädelbach, ed., *Rationalität*, Frankfurt/M.: Suhrkamp, 1984.

Schnelle, Thomas, *Ludwik Fleck – Leben und Denken: Zur Entstehung und Entwicklung des soziologischen Denkstils in der Wissenschaftsphilosophie*, Freiburg: Hochschulverlag, 1982.

Sellars, Wilfrid, "Empiricism and the Philosophy of Mind", in: *Science, Perception and Reality*, London: Routledge & Kegan Paul, 1963.

Shils, E., "Tradition and Liberty: Antinomy and Interdependence", *Ethics* LXVIII/3, April 1958.

Shils, E., "Intellectuals, Tradition, and the Traditions of Intellectuals: Some Preliminary Considerations", *Daedalus*, Spring 1972.

Shils, E., *Tradition*, London: Faber and Faber, 1981.

Simon, Herbert A., "The Consequences of Computers for Centralization and Decentralization", in M. L. Dertouzos and Joel Moses, eds., *The Computer Age: A Twenty-Year View*, Cambridge, Mass.: MIT Press, 1979.

Slobin, Dan I., and Thomas G. Bever, "Children use canonical sentence schemas: A crosslinguistic study of word order and inflections", *Cognition* 12 (1982).

Sluga, Hans, "Thinking as Writing", *Grazer Philosophische Studien*, vol. 33/34 (1989).

Smith, Barry, "Austrian Origins of Logical Positivism", in: B. Gower (ed.), *Logical Positivism in Perspective*, London: Croom Helm, 1987, pp. 35–68.

Smith, Barry, "Practices of Art", in: J. C. Nyíri and B. Smith (eds.), *Practical Knowledge. Outlines of a Theory of Traditions and Skills*, London: Croom Helm, 1988, pp. 172–209.

Smith, Barry, "Textual Deference", *American Philosophical Quarterly* 28/1, 1991.

Sokolowski, Robert, "Natural and Artificial Intelligence", *Daedalus*, Winter 1988.

Spengler, Oswald, *The Decline of the West*, New York: 1928.

Steiner, George, *Martin Heidegger: Eine Einführung*, München: Carl Hanser Verlag, 1989.

Stock, Wolfgang S., "Die Bedeutung Ludwik Flecks für die Theorie der Wissenschaftsgeschichte", *Grazer Philosophische Studien* 10 (1980), pp. 105–118.

Strawson, P. F., "Review of Wittgenstein's *Philosophical Investigations*" (1954), in: G. Pitcher, ed., *Wittgenstein: The Philosophical Investigations. A Collection of Critical Essays*, Anchor Books, 1966.

Szerdahelyi, István, "Entwicklung des Zeichensystems einer internationalen Sprache: Esperanto", in István Fodor – Claude Hagège, eds., *Language Reform: History and Future*, Hamburg: Buske Verlag, 1983/84, vol. 3.

Tauli, Valter, "The Theory of Language Planning", in Joshua A. Fishman, ed., *Advances in Language Planning*, The Hague: Mouton, 1974.

Toffler, Alvin, *The Third Wave*, New York: Bantam Books, 1980.

Tranøy, K. E., "Wittgenstein in Cambridge 1949–51. Some Personal Recollections", in: *Essays on Wittgenstein in Honour of G. H. von Wright – Acta Philosophica Fennica* 28/1–3, 1976.

Turkle, Sherry, *The Second Self: Computers and the Human Spirit*, London: Granada, 1984.

Vygotsky, Lev Semenovich, *Thought and Language* (1934), MIT, 1962.

Wagner, Richard, "Das Judentum in der Musik" (1850), in: Wagner, *Gesammelte Schriften und Dichtungen in zehn Bänden* (ed. W. Golther), Berlin: Deutsches Verlagshaus, n.d., vol. V.

Waldenfels, Bernhard, "The Ruled and the Unruly: Functions and Limits of Institutional Regulations", *Graduate Faculty Philosophy Journal* vol. 9, no.1 (Winter 1982).

Waldenfels, Bernhard, "Die Herkunft der Normen aus der Lebenswelt", in his *In den Netzen der Lebenswelt*, Frankfurt/M.: Suhrkamp, 1985.

Weber, Max, "Die protestantische Ethik und der 'Geist' des Kapitalismus", *Archiv für Sozialwissenschaft und Sozialpolitik* XX–XXI (1903–1904).

Weber, Max, *Economy and Society*, New York: Bedminster Press, 1968.

Weininger, Otto, *Geschlecht und Charakter: Eine prinzipielle Untersuchung*, 25th ed., Wien: Braumüller, 1923.

Wieland, Wolfgang, *Platon und die Formen des Wissens*, Göttingen: Vandenhoeck & Ruprecht, 1982.

Wilkins, Burleigh Taylor, *The Problem of Burke's Political Philosophy*, Oxford: Clarendon Press, 1967.

Winograd, T., and F. Flores, *Understanding Computers and Cognition: A New Foundation for Design*, Norwood, N.J.: Ablex, 1986.

Wittgenstein, Ludwig, *Philosophical Investigations*, Oxford: Basil Blackwell, 1953.

Wittgenstein, Ludwig, *The Blue and Brown Books*, Oxford: Basil Blackwell, 1958.

Wittgenstein, Ludwig, *Zettel*, Oxford: Basil Blackwell, 1967.

Wittgenstein, Ludwig, "Bemerkungen über Frazers *The Golden Bough*", *Synthese*, vol. 17, 1967.

Wittgenstein, Ludwig, *Gespräche, aufgezeichnet von Friedrich Waismann* (Wittgenstein, *Schriften*, vol. 3), Frankfurt am Main: Suhrkamp, 1967.

Wittgenstein, Ludwig, *Lectures and Conversations on Aesthetics, Psychology and Religious Belief*, ed. by Cyril Barrett, Berkeley: University of California Press, 1967.

Wittgenstein, Ludwig, *On Certainty*, Oxford: Basil Blackwell, 1969.

Wittgenstein, Ludwig, *Letters to Russell, Keynes and Moore*, Oxford: Basil Blackwell, 1974.

Wittgenstein, Ludwig, *Philosophical Grammar*, Oxford: Basil Blackwell, 1974.

Wittgenstein, Ludwig, *Bemerkungen über die Grundlagen der Mathematik*, Frankfurt/M.: Suhrkamp, 1974.

Wittgenstein, Ludwig, *Philosophical Remarks*, Oxford: Basil Blackwell, 1975.

Wittgenstein, Ludwig, *Culture and Value*, transl. by Peter Winch, Oxford: Basil Blackwell, 1980.

Wittgenstein, Ludwig, *Wörterbuch für Volksschulen*, ed. by A. Hübner *et al.*, Wien: Hölder-Pichler-Tempsky, 1977.

Wright, G. H. von, "The Wittgenstein Papers", *The Philosophical Review*, vol. 78 (1969), pp. 483–503.

Wright, G. H. von, "Wittgenstein in Relation to his Times", *Proceedings of the 2nd International Wittgenstein Symposium*, Wien: Hölder-Pichler-Tempsky, 1978.

Wright, G. H. von, *Wittgenstein*, Oxford: Basil Blackwell, 1982.

Ziman, John, *Public Knowledge: An Essay concerning the Social Dimension of Science*, Cambridge: Cambridge University Press, 1968.

INDEX

Acham, Karl, 132
Acton, H. B. 62, 141, 159
Ajdukiewicz, K. 123
alienation 37, 42f.
  information deficit as a source of 44
  from the real life-world, effected by
    writing 79
Alloni-Fainberg, Y. 87f., 147f.
Allport, Gordon W. 137f.
Ambrose, Alice 4, 116, 159
anthropology
  conservative 16, 117
  liberal 5
  and morals, Hayek on 134
  traditionalist 5
  of cultural universals 92
Apel, Karl-Otto 93–95, 150, 159
Aristotle 97, 143, 152
Armstrong, D. M. 63, 73, 136, 141
artificial intelligence (AI) 80, 82, 96,
  103, 146, 154
Augustine 14, 140f.
Austria 115, 124
Austrian philosophy 26f., 38, 58f., 68,
  99, 125, 133
authority 2, 31f., 37, 55, 63f.
  of the Bible 61
  and language reform 87, 89, 147f.
  Polanyi on 47, 55
  rule and 68, 138
  of written text 103, 139
autism 156

Baeck, Leo 24, 37, 119, 124f., 159
Baker, G. P. 107f., 121, 154f., 159
Bakhtin, Nikolai 34, 123
Baldamus, W. 125
Bárczi, Géza 87, 147f., 159
Barrett, Cyril 133

Baumbach, R. 135
Bell, Daniel 126f.
Bemporad, J. R. 156
Benedict, Ruth 65, 136, 159
Bennett, Neville 134
Berger, Peter L. 132
Bettelheim, Bruno 156
Bever, Thomas G. 56, 132
Bloch, Joseph 126
Bloomfield, Leonard 56, 91, 147–149,
  159
Bloor, D. 48, 116, 130, 159
Bolinger, Dwight 91, 149, 160
Bolkosky, S. M. 17, 118
Bollnow, O. Fr. 149
Bolter, David 42f., 81, 146
Boltzmann, Ludwig 116
Brandl, J. 151, 154
Brouwer, L. E. J. 10
Brugmann, K. 90, 148, 160
Bruner, Jerome S. 157
Buber, Martin 125
Burke, Edmund 54f., 58f., 67, 132, 138,
  160

Cahnman, Werner J. 124
Calvin, J. 61
Carnap, Rudolf 25, 85, 117, 119, 160
Castañeda, H.-N. 107, 112
Catholicism 36, 38
  and Jewishness 18, 24
  and traditions 61, 135
Chafe, Wallace L. 148f.
Chomsky, Noam 56, 91
Christianity 81, 124f.
  and Jewishness 17, 23, 36
Chwistek, L. 123
Cicero, Marcus Tullius 66
civilization 3, 16, 18f., 56, 60

in the computer age 129
and institutions 137
Clark, Katerina 123
Cohen, Robert S. 120f.
Comte, Auguste 123
Coleman, Samuel 131
communication 37, 48, 52, 54f., 148
  and autistic thinking 156
  global 92
  Hayek on the limits of 51
  Heidegger on, 152
  oral-aural ix, 62, 77, 79, 86, 106,
    108, 111, 155
  and society 75, 142f.
  technology of 44f., 75, 77f.
  written 85, 108, 155f.
computers
  epistemology of 49, 80, 128
  and historical consciousness 75,
    80–83
  and information 44f., 80
  political effects of 44, 129
  computer revolution 39
  sociological effects of 44, 128
  and spoken language 80
  tools vs. machines 42f.
  and word processing 82, 146
consciousness
  autistic 156
  of community 124f.
  historical 73, 77f., 80–83, 145
  machine 128
  of the past 73, 75, 81, 142, 145
  pertaining to tradition 73, 140
  of the worker, Marx on 128
  and writing 77, 81
conservatism 2, 15ff., 24, 82, 112,
    116ff.
  within Esperanto 89
  of written text 139
conventions 55f., 62f., 64f., 67, 69f.,
    91, 134f.
  Hauser on 55f., 58, 69
  Hayek on 126
  Lewis on 136
  Nietzsche on 62, 135f.

in science 48, 119, 129
  Weber on 137, 139
conventional spelling 86
conventionalism 25f., 29f., 121
Coser, Lewis A. 124
creativity 32, 56, 58, 91, 122, 133
culture 26, 30, 35, 83, 88, 142, 145,
    151
  vs. civilization 18f., 56, 60
  computer 42, 82
  liberal x
  literate 80, 112, 145, 153, 157
  manuscript 77
  and morals, Hayek on 134
  oral 75f., 79, 81, 85f., 108f., 111,
    113, 143, 157
  pre/post-literate 81, 85, 105
  typographical 77, 80–82, 112
customs 55f., 59, 62–66, 68–74, 101f.,
    135–137, 139, 141, 142f.
  Burke on 54
  and habits 65f., 69, 97, 101
  and rituals 65
  transmitting knowledge 33, 47
  Weber on 136f.

Deneffe, August, SJ 134, 139
Dertouzos, Michael L. 129
Descartes, René 95, 105
Deutsch, Karl W. 60, 134, 143
Dewey, John 66f., 71, 142, 160
Dijkstra, Edsger W. 83, 146
Dimler, G. Richard, SJ 82, 146
Dostoevsky, F. M. 2, 16, 118
Douglas, Mary 125
Dretske, Fred I. 126
Dreyfus, Hubert L. 96, 150, 161
Dryden, John 137
Duhem, P. M. M. 25
Durkheim, Emile 27–29, 31, 34–36, 68,
    120f., 123–125, 136, 138, 142,
    161

Eastern Europe ix, 39
  liberalization in 40
  political culture of 39f.

Eisenberg, L. 156
Eisenstadt, S. N. 126, 161
Eisenstein, Elizabeth 77f., 145
Eliot, T. S. 54, 132
elites 30, 76, 83
  epistemological significance of 88f.,
    148
  traditional 126
Engelmann, Paul 2, 115, 161
Engels, Friedrich 39, 126f.
Enlightenment 59, 79, 96, 138
epistemology, epistemological 25, 27,
    29–31, 38, 47, 51, 54, 79, 99,
    105, 109, 111, 123, 139, cf. also
    theory of knowledge
Epstein, Klaus 117
Esperanto 89f., 92, 147f.
  Wittgenstein on 85f., 89

Fann, K. T. 116f., 147
Fehér, István 150f.
Feigenbaum, Edward A. 131
Ferguson, C. A. 88, 148
Feyerabend, Paul 48, 53, 130f., 161
Finley, M. I. 144
Fleck, Ludwik 26, 28–34, 49, 55,
    121–123, 125, 130–132, 161
Flores, F. 154
Fodor, J.A. 149
Frank, Tibor 146
Frazer, Sir James 14f., 22, 116f.
Frege, Gottlob 10, 99
Freud, Sigmund 123, 151
Frisé, Adolf 133

Gadamer, Hans-Georg 59, 138f., 145
Garver, Newton 92, 149, 161
Gehlen, Arnold 131
Gellner, Ernest 134
Gethmann, C. F. 94, 150
Giedymin, Jerzy 121, 123
Goethe, J. W. von 135
Gödel, K. 2
Gombocz, W. x, 149
Goodstein, R. L. 116, 161
Goody, Jack 76, 79, 143–145, 152, 156f.

Grassl, Wolfgang 133
Greece 100, 128
Greeks 23, 83, 101, 144
  Heidegger on 101–103
Greek alphabetic writing 76
Greenberg, Joseph H. 88, 147
Grimm brothers 162
  explaining "tradition" 135
Groningen, B.A. van 144
Gumplowicz, Ludwig 27, 34f., 120,
    123f., 162

habit 31, 55–57, 59, 62–71, 91, 139,
    143, 147
  linguistic 91, 147, 149, 154, 156
  and mentality 70
Hacker, P.M.S. 107f., 121, 154f., 159
Haefner, Klaus 79, 128f., 145, 147
Hahn, F. 139
Hajnal, István 79, 129, 144f., 162
Halbwachs, Maurice 28–34, 53, 55f.,
    69, 73, 120–122, 132, 138,
    141–144, 157, 162
Hall, R.A. 86, 89, 147f.
Haller, Rudolf x–xi, 25f., 92, 117,
    119f., 130, 149, 151, 154, 162
Hamilton, Sir William 67
Hart, H. L. A. 59, 69, 133
Haugen, Einar 88f., 147f., 162
Hauser, Arnold 55f., 58, 69, 132, 162
Havelock, Eric 76, 143, 151, 153f., 162
Hayek, F. A. von 40, 51, 59, 126, 131,
    133f., 163
Hebel, J. P. 21, 96, 118, 151
Heidegger, Martin 93–99, 101–103,
    118, 146, 150–154, 163
  influencing AI research 154
  and conservatism 96f., 102
  and everyday language 94
  spirit of orality 101–103, 154
  on the impact of radio 154
  on tradition 96, 153
Hempel, Carl G. 119
Heraclitus, 151, 153
Herrmann, F. W. von 101, 152
Hertz, Heinrich 116

Hertzberg, Lars 154
Hesse, Mary 48
Heyse, K. W. L. 91, 149
Hintikka, Jaakko and Merrill B. 112,
    156f., 163
Hobsbawm, Eric 141
Hockett, Charles F. 85, 91, 147, 149
Holenstein, Elmar 148
Hollinger, David 57, 133
Holquist, Michael 123
Homer 143
Hooker, Richard 61
Humboldt, Wilhelm von 149
Hume, David 55, 152
Hungary
    language reform in the 19th century
        87
    without traditions ix
Husserl, Edmund 37

individuals
    shaped by their community 65, 106,
        157
    in the computer age 44
    as bearers of habits 64f., 70
    literate 108f.
    in a non-literate society 76
    and oral communication 79, 101f.,
        108
    Lyotard on individuals and society
        142
individual memory 28, 108, 157
individual thinking/behaviour ix, 26f.
    Cf. also: 32 (individual creativity, in-
    dividual idiosyncrasies), 33 (indivi-
    dual and collective knowledge), 35
    and 37 (Durkheim: individual
    rooted in its group), 36f. (individual
    reason), 37 (individual intentional
    acts), 38 (collective reason constitu-
    tive of individual minds), 68 (rule as
    governing individual conduct), 69
    (Durkheim: rule is withdrawn from
    individual discretion), 110 (indivi-
    dualized activity), 113 (knowledge
    acquired in an individual way; in-

dividual reading), 79f. (individual
    not fully participating in the literate
    repertoire), 123 (Fleck on Durkheim
    on the individual determinated by so-
    cial structure), 125 (Durkheim vs.
    Weber), 155 (individual deciding on
    A = A), 156 (individualized
    meanings in autistic thinking), 157
    (Goody and Watt on private
    thought)
individual scientist 25f., 29, 57, 119f.
individualism
    Hayek on 126, 134
    philosophical 99
    and universal language 92
individualistic ix
    Gumplowicz on individualistic
        psychology 27
individuality
    all-sided, Marx on 42, 127
    Buber on the dialogical structure of
        37
    and freedom, Vygotsky's relevance
        on 157
information 43
    age of 42, 75, 80-82
    Bell on the economics of 127
    explosion 79, 83, 129
    and knowledge 126f.
    and social cohesion 75, 142f.
    technology, social-political conse-
        quences of 41, 44f., 82f., 128
    embedded in traditions 52, 62, 141
informational deficit 79f.
institutions 47, 57, 63–65, 67, 73, 125,
    137
    Burke on 59
    as embodied habits, Dewey on 67
    devoid of memory 74
    Menger on 59
    liberal 41
    stable ix
Innis, Harold A. 143

Janik, Allan 151
Jerusalem, Wilhelm 34, 37, 123

Jewishness 17–21, 23f., 35–38, 118, 124f., 139
Joas, Hans 120
Johnson, Samuel 89, 148, 164
  definitions by 61, 63–65, 137
Johnston, William M. 124
Jones, O.R. 157
Judaism
  Durkheim on 36
  epistemological message of 38
Jünger, Ernst 18, 118

Käsler, Dirk 124
Kafka, Franz 112, 157
Kaltenbrunner, Gerd-Klaus 117, 164
Kanner, Leo 156
Kant, Immanuel 50, 152
Katz, J. J. 149
Kaulla, Rudolf 19, 164
Kenny, Anthony 117, 164
Keynes, J. M. 115
King, M. D. 130
Klemperer, Klemens von 118
knowledge
  book knowledge 31f., 83
  collective 33, 113
  critical, analytic 77, 81
  economics of 40, 127
  fragmentation of 42f., 44, 79
  habitual 63
  of law and morality, Gadamer on 59
  and myth 76f.
  conveyed by oral traditions 44, 62, 76, 101, 143
  "personal" 49
  Plato on 101, 130
  practical 47–52, 79f., 96, 130
  preserving of 75f.
  "pretheoretical", Berger and Luckmann on 132
  Scheler on 130f.
  scientific 26, 82, 97
  and social factors 26, 30, cf. also sociology of knowledge, theory of knowledge
  tacit 47–49

transmitted by tradition 47
and vision, Heidegger on 153
embodied in writing 44f.
Koselleck, Reinhart 77f., 82, 144f.
König, Gert x
König, René 124
Kraus, Karl 116
Kripke, Saul 108, 154f., 164
Kuhn, Thomas S. 25f., 35, 47, 49, 52, 57–59, 120, 130, 132f., 164
Kümmel, W.G. 139

labour theory of value 40f.
language
  agreement in 48, 113
  artificial/international/universal 85f., 89, 92, 148
  autistic 156
  classical 83
  and community 30, 88, 107, 148, 157
  conventional nature of 149
  fixed through custom 86
  egocentric 110
  factual knowledge pertaining to 91
  formulaic 111, 156
  fragmentation 85, 90
  idiomacity belonging to the essence of 90, 149
  inner 32, 110, 155
  as institution 67
  intersubjective 25, 95
  and meaning 31, 113
  and memory 90f., 109
  mentalist view of 53f.
  philosophy of 112
  printed 81
  spoken 80f., 99–102, 105f., 108f., 111, 153–155
  traditions of 56, 87, 96
  written 77, 81, 86, 99f., 102f., 105f., 108f., 110–112
language reform 87f., 90
language standardization 85–89
Lazerowitz, M. 116, 159
Le Bon, Gustave 123

Leibniz, G. W. 152
Leiser, Burton M. 56, 65, 132
Leoni, Bruno 139
Leskien, August 148, 160
Lessing, G. E. 135
Levi, F. H. 139
Levi, A. W. 116, 165
Levy, David 128
Lévy-Bruhl, L. 34
Lewis, David K. 136
literacy
    emergence of 44, 75f.
    fully unfolded 79f., 108, 113, 144
    petering out 113
    Plato as the philosopher of 103, 154
    spoken language without the background of 111
    transition from orality to 77, 100, 144
Locke, John 65, 105
Luckmann, Thomas 132
Lübbe, Hermann 81, 141, 165
Lukács, Georg 26, 34, 36, 39, 124, 165
Lukes, Steven 35, 124f.
Luria, Alexander R. 155f.
Luther, Martin 61, 135
Lyotard, Jean-François 142

Mach, Ernst 25, 30, 68, 83, 119, 121, 138, 147
McCagg, William O., Jr. 124
McCorduck, Pamela 131
McDougall, William 123
McGuinness, Brian 118, 151, 157, 165
McLellan, David 127
McLuhan, Marshall 146
Malcolm, Norman 154
Malinowski, Bronislaw 105f., 165
Mann, Thomas 16, 118
Mannheim, Karl 2, 26, 34, 37, 116f., 165
Marx, Karl 26, 34, 39–43, 126–128, 165
Marxism 39–43
Maury, André 115
meaning
    in autistic language 156
    Burke on 54

and context, Malinowski on 105f.
Heidegger on 94–96, 102, 150
name theory of 94, 112
and social environment 33
use theory of 12, 14f., 31, 91, 94, 99f., 111–113
Vygotsky on 111
memory
    in autistic thinking 156
    collective 28, 76, 101, 111
    computer 82
    and formulaic language 111, 156
    Halbwachs on 28, 33, 132, 142
    historical 75
    and application of rules 107f., 154
    social function of 144
    Strawson on 109f.
    verbal 110
Mendelssohn-Bartholdy, Felix 19f., 22
Menger, Carl 58f., 133
mental
    images, processes 15, 32, 77, 94, 99
mentalism 53f., 94, 105, 112
    and Wittgenstein 31, 34, 94, 99
mentality 63, 70
Merkel, Reinhard 150
Merton, Robert K. 26, 34, 120, 166
Mill, J. St. 64
Milton, John 62, 66
modernity
    early modern period 77; the modern age 78; modern man, modern life-world 79; "we moderns" (Nietzsche) 79; modern Hebrew 87; Heidegger and Wittgenstein as foes of modernity 96f.; Wittgenstein in opposition to the modern age 99; Wittgenstein not putting together a book in the modern sense 100; modern historical consciousness 77f., 82f., 145
Moeller van den Bruck, Arthur 16, 118
Moss, Gordon E. 142
Murray, Michael 95, 150f., 166
Musil, Robert 58, 68, 133, 157, 166

nationalism 60, 134
Nestroy, Johann Nepomuk 1, 97
Neurath, Anna 35
Neurath, Marie 35, 120
Neurath, Otto 25f., 34f., 119f., 124, 166
Neurath, Wilhelm 35
Németh, László 147, 166
Nietzsche, Friedrich 16, 62, 79, 135f., 166
Nyíri, J. C. 122, 124, 133, 157, 166

Oakeshott, Michael 47, 52f., 57, 129f., 132
Ogden, C. K. 105, 154
Ong, Walter J., SJ 80f., 142f., 145f., 151f., 157, 166
orality
    and Esperanto 147
    Heidegger and Wittgenstein the philosophers of 103
    of manuscript culture 77
    primary 75, 80, 108
    secondary 75, 80, 145f.
    and relevance of situation 105, 112
    and use theory of meaning 99f.

Park, Robert E. 142
Parmenides 97
Pascal, Fania 2, 5, 115, 117, 166
philosophy/theory of science 38, 55
    sociological turn of 26
    traditionalist 49
Piaget, J. 110
Pieper, Josef 73, 140
Plato 79, 97, 99f., 102f., 130, 152
Platonism 96, 99, 154
Plumb, J. H. 145
Pocock, J. G. A. 67, 73, 137, 140, 166
Pöggeler, O. 150
Poincaré, Henri 25
Polanyi, Michael 47–49, 51, 55, 129, 132, 166f.
Popper, Karl R. 37, 52, 57f., 131, 133, 167
    on Jewish assimilation 125
postmodern

view of social cohesion 75
style of Wittgenstein's writings 147
prejudice 55, 62f., 67f., 70, 139
    Allport on 137f.
    Burke on 67f., 138
    Gadamer on 138
    Gumplowicz on 27
printing (typography) 75
    epistemic relevance of 112, 144, 146
    and the coming of modernity 77f.
    Wittgenstein's relation to 100, 151
private language, 105, 107f., 110–113, 154, 156, cf. also Wittgenstein
programming 41–43, 81, 83, 128
progress
    bourgeois-liberal 19
    technological 40, 78, 127
    idea of 1, 77f., 81
    and tradition 59
Protestantism 24
    and printing 145
    and reading 72

Radin, Max 62, 67, 73, 139, 167
radio 80, 85, 103, 145
    Heidegger on 146, 154
Raleigh, Sir Walter 64
Rappaport, Roy A. 143
rationalism 23
    Menger on 59
    Kuhn on 130
rationality
    embedded in social activity 131f.
    Halbwachs on 28f., 53, 132
    and language reform 87
    Oakeshott on 53
    and traditionalism 52f.
    rational standards of spelling 86f.
reason (cf. also rationality)
    collective 38
    Feyerabend on 131
    Gumplowicz on social embeddedness of 27
    individual 27, 36f.
    and tradition 29, 52f., 70f., 73, 130,

132, 140
reflection
    and practical knowledge 47
    socially determined 71
    made possible by the written text
        73, 77, 105f., 111, 139
regress, infinite 31, 51, 141
Renan, E. 22f., 119
Rentsch, Thomas 95, 167
Rhees, Rush 107, 154
Richards, I. A. 105, 154
ritual
    creating cognitive distance 111, 156
    and homeostasis 143f.
    memory reinforced by 76
Rodi, Frithjof 150
Rösler, Wolfgang 143
rules 55, 68, 90f., 96, 154
    constitutive 56
    and convention 64
    and customs 65, 70, 137
    Feyerabend on 48
    handing down of 62
    Hart on 59, 69
    Hayek on 51, 59, 134
    and heuristics 50
    and institutions 67, 137
    of law and morality, Gadamer on 59
    and rituals 136
    of scientific discovery 47
    of spelling 86
    and technical knowledge 130
rule-following 50f., 106, 154
Russell, Bertrand 116
Rutte, Heiner 120
Ryle, Gilbert 47, 50–52, 54, 95, 129, 150

Sampson, Geoffrey 132
Savigny, Eike von 92, 149
Savigny, F. K. v. 58
Schaub, Edward L. 120
Scheler, Max 26, 34, 36, 131, 167
Schlick, Moritz 3, 151
Schmidt, Conrad 126
Schnädelbach, H. 132
Schnelle, Thomas 120f., 123, 160

Schopenhauer, Arthur 135, 152
Schütz, Alfred 37
Schwemmer, Oswald 131f.
Sellars, Wilfrid 109, 155
Shakespeare, W. 66
Shanker, Stuart x
Shils, E. 67, 73, 137, 140, 142, 167
Simmel, Georg 26, 34–36, 123f.
Simon, Herbert A. 129
skills 48f., 64, 70, 79, 83, 96, 139
    scientific, Fleck on 30
Slobin, Dan I. 56, 132
Sluga, Hans 147
Smith, Adam 128
Smith, Barry x, 122, 125, 133, 144, 168
sociology of knowledge (Wissenssozio-
        logie) 26f., 37, 120f.
Sokolowski, Robert 146
Spengler, Oswald 2f., 16, 19–21, 81,
        116, 118
    influence on Fleck 123
    on writing 77, 105
Spinoza, Baruch 152
Sraffa, P. 116, 122, 151
Steiner, George 93, 150
Stock, Wolfgang S. 125
Strawson, P. F. 107, 109, 155
Szerdahelyi, István 148

Tauli, Valter 147
Tennyson, Alfred Lord 64
theory of knowledge
    sociological 25–29, 31, 34, 36, 120
Toffler, Alvin 44, 128f.
Tolstoy, L. 2, 112
tradition
    in art 57f., 133
    and authority 55, 61, 63, 130, 135,
        140
    as collective memory 73, 76, 132,
        139f., 144
    and communication 48, 52, 61f.,
        142
    and convention 55f., 62, 69, 134f.
    and creativity 133
    and custom 62f., 65, 135, 139, 141

definition of 55f., 61–63, 73, 134f., 140f.
democratic 39
Eliot on 54
Engels on 126
Feyerabend on 48, 53
fictitious 73
Fleck on 26, 32, 121
function of 48, 52, 60, 76, 142
Halbwachs on 29, 53, 55
Hayek on 40, 59, 126
and institution 47, 63, 65, 67, 73f., 140, 142
invented 60, 141
and Jewishness 19, 35
Kuhn on 57, 59, 130, 133
and language 56, 58, 73
Mach on 119
and modernization 126
national 65
Nietzsche on 62, 135
Oakeshott on 47
oral 44, 61f., 72, 76, 101, 135, 139, 143, 145
not mirroring the past 76
Polanyi on 47, 55
Popper on 52f.
and practical knowledge 47–49, 52
primary/secondary 52, 59
rendered dysfunctional by printing 78
and reason 3, 29, 52f., 131f.
and reflection 70–73
and ritual 63, 141
and rules 63, 140f.
in science 57, 59, 130
and skill 62f.
Spengler on lack of 19, 116
Weber on 139
written 72, 139
traditionalism ix, 52, 60
epistemological 16
and mentalism 53f.
weak/strong 52f., 55f.
Tr:anøy, K. E. 118
Turing, Alan 128, 146

Turkle, Sherry 42–44, 129
Twaddel, W. F. 149
Twardowski, K. 123

Vygotsky, L. S. 110f., 155, 157, 168

Wagner, Richard 18, 20, 118
Waldenfels, Bernhard 67, 136f., 168
Watt, Ian 76, 79, 143–145, 152, 156f.
Weber, Max 37, 39, 125, 136f., 139, 168
Webster, Noah, 148
definitions by 62, 64–66, 138
Weininger, Otto 18, 20–22, 118
Wiclif, John 61
Wieland, Wolfgang 168
Wilkins, B. T. 132
Winograd, T. 154
Wittgenstein, Ludwig x, 1–7, 9–17, 19–24, 60, 85f., 89, 93–103, 105–109, 111–113, 117–119, 121f., 128, 130, 132–134, 139, 147, 150–157, 169
on authority 2, 5f., 15, 55
and Catholicism 24
on civilization 1, 17, 20, 22f., 60, 97, 151
on preconditions of communication 5f., 113
and conservatism 1f., 9, 15ff., 24, 96f., 102, 112, 116, 118
on creativity 32, 58
on culture 1, 3, 17, 23, 33, 151
on customs 3, 24, 97, 121
film and radio 1, 103, 153
on individuality and the individual 3, 7, 15, 17, 31
on institutions 17, 97, 121
and Jewishness 17, 19–24, 36, 118f.
on agreement in language 6, 31–33, 113
and everyday language 4, 7, 12, 15, 85, 94
on language as given 6f., 12, 14
on language and thought 11, 32, 97f.

and Marxism 1f.
on meaning 10–12, 14f., 31, 33, 91,
    99f., 107
on memory 4, 13, 155
on the mental 4, 13, 15, 32, 155
oral bias of 85f., 100, 103, 106, 108,
    112, 155
emergence of later philosophy 9ff.,
    34, 99
and practical knowledge 48, 52, 54
on private language/thought 11, 13,
    15, 32, 95, 105ff., 111ff.
on progress 1, 97f.
and Protestantism 24
on reason and reflection 1, 3, 5, 15
research on x, 115
on rules and rule-following 5f., 11,
    31f., 51, 58, 107, 121, 149
and Russia 2f., 115
silence and the unsayable 4, 16, 94,
    150

sociological approach 28, 29–31,
    33f.
style 11f.
traditions, traditionalism 1–3, 5–7,
    16, 48, 55, 85f., 97, 112, 116
Wright, G. H. von x, 19, 48, 115f., 118,
    130, 151f., 169

Ziman, John 129

writing
    epistemological significance of ix,
        72f., 76f., 81, 100, 105f., 109f.,
        139, 154, 156
    giving rise to critical-rational
        thought 76f., 105, 110–113, 157
    and language standardization 85f.
    intellectualist prejudice created by
        79
    rudimentary forms of 75f.
    Vygotsky on 110f.